AF392010

Progress in Numerical Simulation for Microelectronics
Vol. 2

Edited by
K. Merten, Siemens AG, Munich
A. Gilg, Siemens AG, Munich

Chris R. Kleijn
Christoph Werner

Modeling of Chemical Vapor Deposition of Tungsten Films

1993

Springer Basel AG

Authors' addresses:

Dr.ir. Chris R. Kleijn
Delft University of Technology
Kramers Laboratorium voor
Fysische Technologie
Prins Bernhardlaan 6
NL-2628 BW Delft, The Netherlands

Dr. Christoph Werner
Siemens AG
ZFE BT ACM 31
Otto-Hahn-Ring 6
D-W-8000 München 83

A CIP catalogue record for this book is available from the Library of Congress, Washington D.C., USA

Deutsche Bibliothek Cataloging-in-Publication Data
Kleijn, Chris R.:
Modeling of chemical vapor deposition of Tungsten films / Chris R. Kleijn;
Christoph Werner. - Basel; Berlin; Boston:
Birkhäuser, 1993
  (Progress in numerical simulation for microelectronics; Vol. 2)
  ISBN 978-3-0348-7743-5   ISBN 978-3-0348-7741-1 (eBook)
  DOI 10.1007/978-3-0348-7741-1

NE: Werner, Christoph:; GT

© 1993 Springer Basel AG
Originally published by Birkhäuser Verlag Basel in 1993.
Softcover reprint of the hardcover 1st edition 1993
Cover Design: Albert Gomm, Basel

PREFACE

Semiconductor equipment modeling has in recent years become a field of great interest, because it offers the potential to support development and optimization of manufacturing equipment and hence reduce the cost and improve the quality of the reactors.

This book is the result of two parallel lines of research dealing with the same subject - Modeling of Tungsten CVD processes -, which were performed independently under very different boundary conditions. On the one side, Chris Kleijn, working in an academic research environment, was able to go deep enough into the subject to lay a solid foundation and prove the validity of all the assumptions made in his work. On the other side, Christoph Werner, working in the context of an industrial research lab, was able to closely interact with manufacturing and development engineers in a modern submicron semiconductor processing line.

Because of these different approaches, the informal collaboration during the course of the projects proved to be extremely helpful to both sides, even though - or perhaps because - different computer codes, different CVD reactors and also slightly different models were used. In spite of the inconsistencies which might arise from this double approach, we feel that the presentation of both sets of results in one book will be very useful for people working in similar projects.

We greatfully acknowledge the opportunity given by the editors Knut Merten and Albert Gilg to publish this book in their series "Progress of Numerical Simulation in Microelectronics (PNSM)" in the Birkhäuser Verlag. We would also like to acknowledge the cooperation of the people of the Siemens equipment simulation group, R.P Brinkmann, P. Flynn, Chr. Hopfmann, A. Kersch, and especially with J.I.Ulacia F., who contributed to the Siemens project with his profound knowledge in Chemistry, Computer Science and Semiconductor Technology. Special thanks are also due to H. Körner in the Siemens processing line for providing experimental data as well as many helpful comments.

Furthermore, we would like to thank C.J. Hoogendoorn and Th.H. van der Meer of Delft University of Technology for their valuable suggestions and G.C.A.M. Janssen, C. van der Jeugd and G.J. Leusink of the Delft Institute for Micro Electronics and Submicron Technology for many fruitful discussions. Finally, we like to give special credits to A. Hasper and J. Holleman from the MESA Institute for Microelectronics of Twente University. Through many fruitful discussions and valuable suggestions and by making available their experimental data on blanket tungsten CVD they contributed largely to the research presented in chapter 4 of this book.

Many thanks are also due to M.D. Mossel for typing large parts of the manuscript, and to Bill Morokoff for his critical reading of the manuscript.

Contents

PRINCIPAL SYMBOLS

a	radiative absorptivity
A	surface area, m^2
c	mole concentration ($=P/RT$ for ideal gas), $mole \cdot m^{-3}$
c_p	specific heat of the gas mixture, $J \cdot kg^{-1} \cdot K^{-1}$
d	distance between wafer and susceptor, m
D	binary ordinary diffusion coefficient, $m^2 \cdot s^{-1}$
$\mathbb{D}$	effective multicomponent diffusion coefficient, $m^2 \cdot s^{-1}$
$\mathbb{D}^T$	multicomponent thermal diffusion coefficient, $kg \cdot m^{-1} \cdot s^{-1}$
$\mathbb{D}^K$	Knudsen diffusion coefficient, $m^2 \cdot s^{-1}$
e	radiative emissivity
E_A	activation energy, $kJ \cdot mole^{-1}$
f	species mole fraction
F	radiative viewfactor
$\underline{g}$	gravity vector ($g_z = -9.81\ m \cdot s^{-2}$)
G_i^0	standard Gibbs energy change of formation for the ith species, $kJ \cdot mole^{-1}$
ΔG_k^0	standard Gibbs energy change for the kth reaction, $kJ \cdot mole^{-1}$
G_{ij}	Gebhardt factor (chapter 3)
$\mathcal{G}$	growthrate, $m \cdot s^{-1}$
H^0	standard heat of formation, $J \cdot mole^{-1}$
H	molar enthalpy, $J \cdot mole^{-1}$
$\underline{\underline{I}}$	unity tensor
$\underline{j}$	diffusive mass flux vector, $kg \cdot m^{-2} \cdot s^{-1}$
$\underline{J}$	diffusive mole flux vector, $mole \cdot m^{-2} \cdot s^{-1}$
J_q	heat flux, $W \cdot m^{-2}$
k_k	forward reaction rate constant for the kth homogeneous reaction, in m^3, $mole$ and s, depending on the reaction order
k_{-k}	reverse reaction rate constant for the kth homogeneous reaction, in m^3, $mole$ and s, depending on the reaction order
K	number of gas phase reactions
K_k	equilibrium constant for the kth gas phase reaction
ℓ	mean free path length, m
L	number of surface reactions
L_0	initial depth of trench or contact hole, m
$\mathcal{L}$	characteristic reactor dimension, m
m	average mole mass, $kg \cdot mole^{-1}$
m_i	mole mass of the ith species, $kg \cdot mole^{-1}$
M	number of surface species

$\underline{n}$	unity vector normal to the inflow/outflow opening or wall
n	nucleation rate, $s^{-1} \cdot m^{-2}$
N	number of gaseous species
N_A	Avogadro's number ($6.024 \cdot 10^{23}\ mole^{-1}$)
P	pressure, Pa
P^0	standard pressure $= 1.013 \cdot 10^5\ Pa$
Q	volume flow rate at standard conditions, in $slm =$ standard liters per minute ($1\ slm = 6.813 \cdot 10^{-4}\ mole \cdot s^{-1}$ for an ideal gas)
r,z	cylindrical coordinates, m
r	radiative reflectivity
R	universal gas constant $= 8.314\ J \cdot mole \cdot K^{-1}$
$\mathcal{R}_k$	forward reaction rate of the kth gas phase reaction, $mole \cdot m^{-3} \cdot s^{-1}$
$\mathcal{R}_{-k}$	reverse reaction rate of the kth gas phase reaction, $mole \cdot m^{-3} \cdot s^{-1}$
$\mathcal{R}_l^s$	reaction rate for the lth surface reaction, $mole \cdot m^{-2} \cdot s^{-1}$
S^0	standard entropy, $J \cdot mole^{-1} \cdot K^{-1}$
S_ϕ	source term in general transport equation (various units)
t	time, s
T	temperature, K
T^0	standard temperature $= 298.15\ K$
T^*	reduced temperature $= kT/\epsilon$
$\underline{v}$	velocity vector, $m \cdot s^{-1}$
$\mathcal{W}$	trench width or contact hole radius, m
$\|x\|$	$= 0.5(\mid x \mid + x)$
x,y	cartesian coordinates, m

greek symbols

α	thermal accomodation coefficient (chapter 3)
α	thermal diffusion factor (chapters 2 and 5)
β	dimensionless constant defined in section 3.3
γ	reactive sticking coefficient
Γ_ϕ	diffusion coefficient in general transport equation
ϵ/k	ratio of maximum energy of attraction in Lennard-Jones interaction potential and the Boltzmann constant, K
θ	fraction of surface covered with tungsten
κ	volume viscosity, $kg \cdot m^{-1} \cdot s^{-1}$.
λ	thermal conductivity of the gas mixture, $W \cdot m^{-1} \cdot K^{-1}$
μ	dynamic viscosity of the gas mixture, $kg \cdot m^{-1} \cdot s^{-1}$.

ν_{ik}	stoichiometric coefficient for the ith gaseous species in the kth gas phase reaction
ξ_ϕ	coeffient in general transport equation (various units)
ρ	density, $kg \cdot m^{-3}$
σ	collision diameter in Lennard-Jones interaction potential, $\mathring{A}ng$
σ_{il}	stoichiometric coefficient for the ith gaseous species in the lth surface reaction
$\underline{\underline{\tau}}$	viscous stress tensor, $N \cdot m^{-2}$
ϕ	general variable in general transport equation (various units)
χ_{jl}	stoichiometric coefficient for the jth surface species in the lth surface reaction
ω	species mass fraction
Ω_μ	tabulated function of T^* for calculating μ (table 2.5)
Ω_D	tabulated function of T^* for calculating D (table 2.5)

subscripts

c	at a cold surface
h	at a hot surface
i,j	with respect to the ith/j^{th} species
ij	with respect to gas pair $i - j$
k	with respect to the kth gas phase reaction
l	with respect to the lth surface reaction
N,E,S,W	in north, south, east, west neighbour grid cell
n,e,s,w	on north, south, east, west wall of grid cell
x,y	in the x, y direction

superscripts

0	at standard temperature and pressure
C	due to ordinary diffusion
$cond$	due to conductive heat flow
rad	due to radiative heat flow
T	due to thermal diffusion
$\dagger$	transposed vector

Chapter 1

Introduction

1.1 Equipment modeling

Semiconductor processing technology, which is the basis of the modern microelectronic revolution, has seen a tremendous increase in complexity in the last decade. This has lead to an exponential growth in the necessary investment of time and money to develop new generations of Integrated Circuits and to provide the equipment to manufacture them. For instance the investment to build a future 64 Mbit DRAM manufacturing line with 0.35 μm minimum feature size is estimated to be in the order of 1 billion US \$. Most of the cost stems from the development and production of the equipment needed to manufacture the integrated circuits, such as optical projection steppers, CVD reactors, reactive ion etchers, ion implanters etc..

In the past the construction of a piece of equipment has been performed largely on a 'trial and error' base. Existing equipment has been adjusted to meet increased demands concerning homogeneity and throughput in a purely empirical way, relying mainly on the 'feeling' of an experienced engineer. Of course this procedure requires several iteration cycles until a satisfactory performance level can be achieved. If this kind of approach is maintained, it seems very difficult and extremely expensive to meet the severe manufacturing challenges that are associated with device structures and minimum feature sizes projected for the next IC generations.

It is generally felt that the process of developing new equipment can be greatly improved if a simulation tool is used to support the design and optimization phase, similar to what is already utilized in circuit and de-

vice design. Process, device and circuit simulation programs have been
successfully used at least in the development of the last three generations
of integrated circuits (256K, 1M, and 4 Mbit DRAMS) and have already
proven their ability to support the optimization of processing conditions,
device structures, transistor design and layout. The main advantage of the
modeling technique in the development of a new product is the possibility
to determine the electrical characteristics of the circuit even before any
part of the device has been fabricated. In reality, experiments and simulation are done simultaneously to provide feedback at the design stage.

The deposition of thin films with Chemical Vapor Deposition (CVD) is a
process of growing importance in semiconductor technology. The chemical
and physical processes that determine the performance of a CVD reactor
- though rather complex - are fairly well understood, so that accurate
predictive models for the reactors can be constructed. Already more than
a decade ago the basic work concerning the differential equations and the
respective boundary conditions was established [1.1] and a number of
basic investigations for CVD processes have been published since then.

Equipment simulation of reactors for undoped polysilicon deposition from
SiH_4 gas appears to be the most widely studied [1.2, 1.3, 1.4], and the
models have been validated through comparison with experiments [1.5].
CVD of silicon dioxide [1.6], silicon nitride [1.7], MOCVD for GaAs
deposition [1.8], and tungsten CVD [1.9, 1.10, 1.11, 1.12], have also been
studied through simulation, although the chemical reaction models used
have been less thoroughly investigated.

In general, most of these works seem to study the principal behaviour of
each process, but few proceed as far as to the optimization of a particular
piece of equipment. Through the availability of increased computing power
it has now become possible to use the simulation as a predictive tool in
the design and optimization of equipment and processes which are used
in industrial CVD applications. [1.13, 1.14].

1.2   Multidisciplinary approach

The successful modeling of CVD reactors requires the utilization of quite
a number of different scientific disciplines in a multidisciplinary approach.

- **Transport physics** of ideal gases is required to establish the necessary differential equations and models for the transport parameters,

which describe the motion of a mixture of gases at low pressure, high temperature and relatively low velocities. This discipline seems to be well developed and understood since more than 50 years.

- **Gas phase chemistry** leads to the formation of reactive intermediates, which may play an essential role in the formation of the required thin films. The required theoretical description of homogeneous reaction kinetics (based on statistical thermodynamics, transition state theory and RRKM analysis) is relatively well established. Theoretical modeling may be used in the determination of important reaction paths and for a semiquantitative prediction of reaction kinetics. However, detailed experiments are still necessary in order to obtain accurate rate constants, which may be used in predictive process modeling.

- **Surface chemistry** is required to describe the chemical reactions leading to film deposition in the CVD process. Here the state of the art is rather crude, especially for the more exotic processes used in modern IC manufacturing, and there is a painful lack of both theoretical models and experimental data characterizing the processes. One has to resort to rather simple empirical models which contain several fitting parameters to allow adjustment to a specific experimental situation.

- **Computational fluid dynamics** is the vehicle to solve partial differential equations and obtain numerical results for each specific working condition on a particular reactor. This field has reached an impressive state of the art and is still in rapid development triggered by the ever increasing computation power of modern computers.

- **Semiconductor technology** finally is the engineering field which will benefit from the CVD modeling, and an intimate knowledge of this discipline is mandatory to develop and apply simulation programs for CVD reactors properly. The urgent demand of this technology for better modeling support is the driving force behind today's activities in CVD reactor simulation.

1.3 Thin Tungsten Films

Tungsten is used in modern integrated cicuits as a metallization and contact layer because of its high melting point, low electrical resistivity and high chemical and electrical stability. Moreover the deposition of a tungsten film by a CVD process allows a very good edge coverage and filling of small holes with a high aspect ratio without void formation. The increased interest in tungsten in modern IC technology is reflected by a drastic increase in publications investigating tungsten CVD processes, most of them first published in the annual *'Workshop on Tungsten and other Refractory Metals for VLSI Applications'*. An extensive presentation of the technological status and the potential problems of tungsten CVD processes can be found in the recent book of J.Schmitz [1.15].

1.4   Outline of the book

In this book the present status of CVD modeling for tungsten deposition will be described. Chapter 2 will develop the mathematical framework describing the hydrodynamics, transport phenomena, and homogeneous and heterogenous chemical reactions,including their influence on the properties of the deposited layers, as well as numerical techniques for the solution of the equations appearing in this model. The model is based on fundamental physical and chemical laws and theories, rather than empirical correlations, and therefore generally applicable to a large variety of thermal CVD processes and reactors.

Chapter 3 will be concerned with the calculation of temperature profiles on the solid boundaries of the reactor chamber. We will see that, although the relevant physical laws are well established, the modeling will be far less fundamental than for the heat transport in the gas volume, since the thermal and geometric boundary conditions are less accurately known. An empirical model developed for a special reactor will be shown.

In chapter 4 this general theory is applied to a particular axisymmetric cold wall CVD reactor for tungsten deposition. We investigate the chemical reactions and transport phenomena using the simulation framework developed in chapter 2. Velocity, temperature and concentration profiles are presented and discussed with respect to optimum deposition behaviour. Additional models are developed for the prediction of the step coverage of blanket via filling processes. The model simulations are compared to experimental growth rate and step coverage data.

In chapter 5 we concentrate on selective tungsten deposition. The chemistry model will be enhanced to cover the different deposition behaviour of tungsten on conductive and insulating surfaces, and a nucleation model will be developed that can describe selectivity breakdown.

Finally chapter 6 will summarize the advantages of the simulation technique as well as discuss the basic limitations and future challenges of CVD modeling.

References Chapter 1

[1.1]   G. Wahl (1977) "Hydrodynamic description of CVD processes",
Thin Solid Films **40** (1977) p.13.

[1.2]   K.F Roenigk and K.F. Jensen (1985), "Analysis of
Multicomponent LPCVD Processes",
J. Electrochem. Soc. **132** p.449

[1.3]   M.E. Coltrin, R.J. Kee, and G.H. Evans, (1989)."A Mathematical
Model of the Fluid Mechanics and Gas-Phase Chemistry in a
Rotating Disk CVD Reactor",
J. Electrochem. Soc. **136** p.819

[1.4]   C.R. Kleijn (1991), "A Mathematical Model for the hydrodynamics
and gas-phase reactions in silicon LPCVD in a Single Wafer
Reactor", *J. Electrochem. Soc.* **138**, pp.2190-2200.

[1.5]   Ch. Hopfmann, J. Ignacio Ulacia F., and Ch. Werner (1991),
"3-D Fluid Simulation of a Polysilicon LPCVD Reactor",
Applied Surface Science **52** pp.169-187

[1.6]   S.R. Kalidindi and S.B. Desu (1990), "Analytical Model for the
Low Pressure Chemical Vapor Deposition of SiO_2 from TEOS",
J. Electrochem. Soc. **137** p.624.

[1.7]   K.F Roenigk and K.F. Jensen (1987), "Low Pressure CVD of Silicon
Nitride", *J. Electrochem. Soc.* **134** p1777 (1987).

[1.8]   H. Moffat and K.F. Jensen (1986), "Complex Flow Phenomena in
MOCVD Reactors", *Journal of Crystal Growth* **77** p.108.

[1.9]   J.I. Ulacia F., S. Howell, H. Körner, and Ch. Werner (1989),
"Flow and Reaction Simulation of a Tungsten CVD reactor",
Appl. Surf. Sci. **38** 370.

[1.10] C.R. Kleijn and C.J. Hoogendoorn; A. Hasper, J. Holleman and
J. Middelhoek (1990) "An experimental and modelling study of
the tungsten LPCVD growth kinetics from $H_2 - WF_6$ at low WF_6
partial pressures" In "Tungsten and other advanced metals for

VLSI/ULSI applications V", S. Wong and S. Furukawa (*eds*), The
Materials Research Society, Pittsburgh, USA, pp. 109-116

[1.11] C.R. Kleijn, and C.J. Hoogendoorn; A. Hasper, J. Holleman, J.
Middelhoek (1991) "Transport phenomena in tungsten LPCVD
in a single-wafer reactor", *J. Electrochem. Soc.* **138**, pp. 509-517

[1.12] Chr. Werner, J. Ignacio Ulacia F., Ch. Hopfmann, and P. Flynn,
(1991), "Equipment simulation of selective tungsten deposition"
J. Electrochem. Soc. **139**, pp.566-574.

[1.13] Chr. Werner, J.I. Ulacia F., and S. Howell (1989),
"Numerical Simulation of Gas Flow and Chemical Reactions
in Semiconductor Processing Equipment", Symposium on VLSI
echnology, May 89 in Kyoto Japan, p. 49.

[1.14] J. I. Ulacia F.,Chr.Werner (1990), "Equipment Simulation",
Solid State Technology ,October 1990, p.47

[1.15] John E.J.Schmitz (1991), Chemical Vapor Deposition of Tungsten
and Tungsten Silicides, Noyes Publ., Park Ridge, New Jersey.

Chapter 2

Mathematical models for chemical vapor deposition

2.1   Introduction

A generally applicable CVD simulation model may consist of a set of partial differential equations with appropriate boundary conditions, describing the gas flow, the transport of energy and species and the chemical reactions in the reactor. In these equations, several properties of the gas mixture appear in relation to its temperature, pressure and composition. The chemical reactions in the gas phase and at the wafer surface can be described in a general way, suitable for many different processes. However, for the actual modeling of a particular process knowledge of the chemical mechanisms and kinetics is required, which must be obtained from experiments and/or theoretical considerations on the particular CVD process chemistry.

In section 2.2 some simplifications are described, which are generally valid for CVD processes and which reduce the complexity of a CVD model. In sections 2.3 and 2.4 the partial differential equations describing gas flow, heat and gas species transport are presented. In sections 2.5 and 2.6 we describe the boundary conditions for these equations and in section 2.7 it is described how the properties of gas mixtures can be predicted from kinetic theory of gases. Finally, in sections 2.8 and 2.9 numerical techniques are described for the solution of the CVD model equations.

Table 2.1: Mean free path lengths in CVD

| Pressure | | Mean free path length | |
| | | Nitrogen | Hydrogen |
Torr	*Pa*	*m*	*m*
0.1	$\simeq 13.3$	$1.2 \cdot 10^{-3}$	$1.9 \cdot 10^{-3}$
1	$\simeq 133$	$1.2 \cdot 10^{-4}$	$1.9 \cdot 10^{-4}$
10	$\simeq 1330$	$1.2 \cdot 10^{-5}$	$1.9 \cdot 10^{-5}$
760	$\simeq 10^5$	$1.6 \cdot 10^{-7}$	$2.6 \cdot 10^{-7}$

2.2 Some model simplifications

Some simplifications can be made which largely reduce the complexity of the problem and the computational effort needed for the solution of the modeling equations and which are so generally justified for CVD conditions that they do not essentially limit the accuracy and the applicability of the model.

(i) The gas mixture can be treated as a continuum. This assumption is valid when the mean free path length ℓ of the molecules is much smaller than a characteristic dimension $\mathcal{L}$ of the reactor geometry, *i.e.* when the Knudsen number is small: $Kn = \ell/\mathcal{L} \ll 1$. In general, the continuum approach is valid for $Kn < 0.01$. Approximate values for the mean free path lengths of the molecules in hydrogen and nitrogen gases at 700 K are shown in table 2.1. For pressures > 1 *Torr* and typical dimensions > 0.01 m we may safely use the continuum approach. For very low pressures and small typical dimensions however, we enter the so called slip flow or the transition regime, where the continuum assumption is invalid. For the modeling of CVD processes in these regimes, statistical modeling approaches for the behavior of individual gas molecules may be applied, such as the Direct Simulation Monte Carlo Method developed by Bird [2.1]. Such models will not be dealt with in this book. Some aspects of free molecular flow behavior in CVD processes will be discussed in chapters 3.3 and 4.5.

(ii) For the pressures and temperatures used in CVD the gases may be treated as ideal gases, behaving in accordance with the ideal gas law and Newtons law of viscosity.

(iii) The gas flow can be assumed to be laminar. In general, a flow be-

Table 2.2: The Grashof number in CVD reactors

| Pressure | | Grashof number | |
| | | Nitrogen | Hydrogen |
$Torr$	Pa	m	m
0.1	$\simeq 13.3$	$10^{-4} - 10^{-1}$	$10^{-6} - 10^{-3}$
1	$\simeq 133$	$10^{-2} - 10^{+1}$	$10^{-4} - 10^{-1}$
10	$\simeq 1330$	$10^{+0} - 10^{+3}$	$10^{-2} - 10^{+1}$
760	$\simeq 10^{5}$	$10^{+4} - 10^{+7}$	$10^{+2} - 10^{+5}$

comes turbulent when either the Reynolds number $Re = V\mathcal{L}/\nu$ or the Grashof number $Gr = g\beta\mathcal{L}^3\Delta T/\nu^2$ becomes very large, with V a typical gas velocity, ν the kinematic viscosity of the gas, β its thermal expansion coefficient, g the gravity constant and ΔT the temperature difference between hot and cold walls. For typical CVD processes, $Re \simeq 10^{-2} - 10^2$, depending on the flow rate, the type of gas and the reactor dimensions, but independent of the total pressure. This is well below values at which the onset of turbulence might be expected. For the Grashof number in coldwall CVD reactors we find values as in table 2.2. Turbulent motion in a stagnant horizontal gas layer heated from below will develop at Grashof numbers $> 10^5$, whereas a superimposed forced convection flow as found in CVD reactors has a stabilizing effect, delaying the transition to turbulent motion to even higher Grashof numbers. So, when using hydrogen as a carrier gas the flow will always be laminar. With nitrogen as a carrier gas, the flow in low pressure CVD will be laminar also, but in atmospheric pressure cold wall CVD reactors with large vertical dimensions and heated from below the flow may be turbulent.

(iv) Although some reactant gases in CVD do absorb infrared heat radiation, these gases are usually present in low partial pressures, either because they are highly diluted or because the total pressure is very low. Therefore, the gas mixture in CVD reactors may be treated as transparent for heat radiation from the heated walls and susceptor.

(v) The viscous heating of the gas mixture due to dissipation may be neglected, since no large velocity gradients appear in CVD gas flows.

(vi) The effects of pressure variations on the temperature of the gas mixture may be neglected for the low Mach number flows in CVD reactors.

2.3   Fluid flow and heat transfer

With the previous simplifications, the gas flow in CVD reactors is described by the conservation equation for mass, or continuity equation

$$\frac{\partial \rho}{\partial t} = -\nabla \cdot (\rho \underline{v}) \tag{2.1}$$

and the conservation equation for momentum, or the Navier-Stokes equation

$$\frac{\partial \rho \underline{v}}{\partial t} = -\nabla \cdot (\rho \underline{v}\underline{v}) + \nabla \cdot \underline{\underline{\tau}} - \nabla P + \rho \underline{g} \tag{2.2}$$

with ρ the gas density, t the time, $\underline{v}$ the gas velocity and P the pressure [2.2]. For Newtonian fluids as the gases in CVD reactors, the viscous stress tensor $\underline{\underline{\tau}}$ takes the form

$$\underline{\underline{\tau}} = \mu \left(\nabla \underline{v} + (\nabla \underline{v})^{\dagger} \right) + (\kappa - \frac{2}{3}\mu)(\nabla \cdot \underline{v}) \cdot \underline{\underline{I}} \tag{2.3}$$

where $\mu = \rho \nu$ is the dynamic viscosity of the gas, κ its bulk viscosity and $\dagger$ indicates a transposed vector.

Due to buoyancy effects, these equations are coupled to the energy equation, describing the temperature field and the heat transfer in the reactor:

$$
\begin{aligned}
c_p \frac{\partial \rho T}{\partial t} ={}& -c_p \nabla \cdot (\rho \underline{v} T) + \nabla \cdot (\lambda \nabla T) + \nabla \cdot \left(RT \sum_{i=1}^{N} \frac{\mathbb{D}_i^T}{m_i} \nabla(\ln f_i) \right) \\
& + \sum_{i=1}^{N} \frac{H_i}{m_i} \nabla \cdot \underline{j}_i - \sum_{i=1}^{N} \sum_{k=1}^{K} H_i \nu_{ik} (\mathcal{R}_k^g - \mathcal{R}_{-k}^g)
\end{aligned} \tag{2.4}
$$

Here, c_p is the specific heat per unit mass of the gas, T its temperature, λ its thermal conductivity and R the universal gas constant. The mole fraction of the ith species is indicated by f_i, its molar mass by m_i, its thermal diffusion coefficient by $\mathbb{D}_i^T$, its molar enthalpy by H_i and its total diffusive mass flux by $\underline{j}_i$. Finally, the stoichiometric coefficient of the ith species in the kth gas phase reaction (with forward reaction rate $\mathcal{R}_k^g$ and reverse reaction rate $\mathcal{R}_{-k}^g$) is written as ν_{ik}.

The third term on the right-hand side of *eq.* 2.4 represents the Dufour effect (or diffusion-thermo effect), causing an energy flux as a result of concentration gradients. The fourth term on the right-hand side represents the transport of heat associated with the interdiffusion of the chemical species. These two effects are probably not too important in most CVD processes. The last term represents the heat production/destruction due to chemical reactions in the gas mixture. This term may be unimportant in processes in which gas phase reactions are negligible, or where the reactants are highly diluted in an inert carrier.

In general, the above equations must be solved in full 3D form to find the gas flow in the reactor. Many single-wafer tungsten CVD reactors however are shaped axisymmetrically, thus reducing the flow to a 2D problem. In cylindrical form, equations 2.1-2.4 can be written as

$$
\frac{\partial}{\partial t}(r\rho) = -\frac{\partial}{\partial r}(r\rho v_r) - \frac{\partial}{\partial z}(r\rho v_z)
$$

$$
\frac{\partial}{\partial t}(r\rho v_r) = -\frac{\partial}{\partial r}(r\rho v_r^2) - \frac{\partial}{\partial z}(r\rho v_r v_z)
$$

$$
+\frac{\partial}{\partial r}\left(2r\mu\frac{\partial v_r}{\partial r}\right) + \frac{\partial}{\partial z}\left(r\mu\frac{\partial v_r}{\partial z}\right) + \frac{\partial}{\partial z}\left(r\mu\frac{\partial v_z}{\partial r}\right)
$$

$$
+r\frac{\partial}{\partial r}\left((\kappa - \frac{2}{3}\mu)(\frac{1}{r}\frac{\partial r v_r}{\partial r} + \frac{\partial v_z}{\partial z})\right) - \frac{2\mu v_r}{r} - r\frac{\partial P}{\partial r}
$$

$$
\frac{\partial}{\partial t}(r\rho v_z) = -\frac{\partial}{\partial r}(r\rho v_r v_z) - \frac{\partial}{\partial z}(r\rho v_z^2)
$$

$$
+\frac{\partial}{\partial z}\left(2r\mu\frac{\partial v_z}{\partial z}\right) + \frac{\partial}{\partial r}\left(r\mu\frac{\partial v_z}{\partial r}\right) + \frac{\partial}{\partial r}\left(r\mu\frac{\partial v_r}{\partial z}\right)
$$

$$
+r\frac{\partial}{\partial z}\left((\kappa - \frac{2}{3}\mu)(\frac{1}{r}\frac{\partial r v_r}{\partial r} + \frac{\partial v_z}{\partial z})\right) - r\frac{\partial P}{\partial z} + r\rho g_z
$$

$$
c_p\frac{\partial}{\partial t}(r\rho T) = -c_p\left(\frac{\partial}{\partial r}(r\rho v_r T) + \frac{\partial}{\partial z}(r\rho v_z T)\right)
$$

$$
+\frac{\partial}{\partial r}\left(\lambda r\frac{\partial T}{\partial r}\right) + \frac{\partial}{\partial z}\left(\lambda r\frac{\partial T}{\partial z}\right)
$$

$$
+\frac{\partial}{\partial r}\left(RTr\sum_{i=1}^{N}\frac{\mathbb{D}_i^T}{m_i f_i}\frac{\partial f_i}{\partial r}\right) + \frac{\partial}{\partial z}\left(RTr\sum_{i=1}^{N}\frac{\mathbb{D}_i^T}{m_i f_i}\frac{\partial f_i}{\partial z}\right)
$$

$$
+\sum_{i=1}^{N}\frac{H_i}{m_i}\left(\frac{\partial r j_{i,r}}{\partial r} + \frac{\partial r j_{i,z}}{\partial z}\right) - \sum_{i=1}^{N}\sum_{k=1}^{K}H_i\nu_{ik}(\mathcal{R}_k^g - \mathcal{R}_{-k}^g)
$$

In the above equations, the fluid properties λ, μ, ρ, $\mathbb{D}_i^T$ and c_p are not only functions of temperature and pressure, but also of the composition of the gas mixture. Thus, the energy and flow equations are coupled to the species concentration equations.

2.4   Species transport and chemical reactions in the gas phase

Gas diffusion in a CVD reactor may result from concentration gradients (ordinary diffusion), but also from temperature gradients (thermal diffusion, Soret effect). There are many different ways of expressing species concentrations and diffusion velocities [2.2, pp. 495-502]. Here we will use mass fractions and diffusive mass fluxes relative to the mass averaged velocity of the gas mixture. The main advantage of this procedure is the fact that the mass averaged velocity is obtained from the Navier-Stokes equations and further also that the resulting convection-diffusion equation is very similar in form to the other transport equations. This is favorable for the numerical solution of the equations.

2.4.1   Species concentration equations

We define the mass averaged velocity $\underline{v}$ in an N component gas mixture as

$$\underline{v} = \sum_{i=1}^{N} \omega_i \underline{v}_i \tag{2.5}$$

and the diffusive mass flux vector $\underline{j}_i$ of the ith species as

$$\underline{j}_i = \rho \omega_i (\underline{v}_i - \underline{v}) \tag{2.6}$$

where ω_i is the mass fraction of the ith species $(i = 1, N)$ and $\underline{v}_i$ the velocity vector of the ith species. We further assume that K reversible chemical reactions take place in the gas phase, with a forward reaction rate $\mathcal{R}_k^g$ $(k=1,K)$, a reverse reaction rate $\mathcal{R}_{-k}^g$ and stoichiometric coefficients ν_{ik}. This is further described in section 2.4.4. Now, the balance equation for the ith gas species, in terms of mass fractions and diffusive mass fluxes, can be written as

$$\frac{\partial \rho \omega_i}{\partial t} = -\nabla \cdot (\rho \underline{v} \omega_i) - \nabla \cdot \underline{j}_i + m_i \sum_{k=1}^{K} \nu_{ik} (\mathcal{R}_k^g - \mathcal{R}_{-k}^g) \tag{2.7}$$

The total diffusive mass flux $\underline{j}_i$ of the ith species is composed of diffusion fluxes due to concentration gradients $\underline{j}_i^C$ (see section 2.4.2) and diffusion fluxes due to thermal diffusion $\underline{j}_i^T$ (section 2.4.3):

$$\underline{j}_i = \underline{j}_i^C + \underline{j}_i^T \tag{2.8}$$

In an N component gas mixture there are $N-1$ independent species concentration equations of the form of equation 2.7, since the mass fractions must sum up to 1:

$$\sum_{i=1}^{N} \omega_i = 1 \tag{2.9}$$

2.4.2 Ordinary diffusion

In a binary gas mixture ($N=2$) the ordinary diffusive mass flux of the species is given by Ficks Law:

$$\underline{j}_1^C = -\underline{j}_2^C = -\rho D_{12} \nabla \omega_1 = \rho D_{12} \nabla \omega_2 \tag{2.10}$$

with D_{12} the binary diffusion coefficient for ordinary diffusion in the pair of gases 1 and 2, which is essentially independent of the mixture composition.

A general expression for the ordinary diffusion fluxes $\underline{j}_i^C$ in a multicomponent ($N > 2$) gas mixture is given by the Stefan-Maxwell equations. They form a set of equations relating the diffusive fluxes of all species in the mixture to all concentration gradients. Written in terms of mole fractions and mole fluxes we have

$$\nabla f_i = \frac{1}{c} \sum_{j=1}^{N} \frac{1}{D_{ij}} \left(f_i \underline{J}_j^C - f_j \underline{J}_i^C \right) \tag{2.11}$$

with f_i the mole fraction of the ith species, c the total mole concentration of the gas mixture ($=P/RT$), D_{ij} the binary diffusion coefficient for the ith and j^{th} species and $\underline{J}_i^C$ ($=\underline{j}_i^C/m_i$) the diffusive mole flux of the ith species. In terms of mass fractions and fluxes we obtain

$$\nabla \omega_i + \omega_i \nabla (\ln m) = \frac{m}{\rho} \sum_{j=1}^{N} \frac{1}{m_j D_{ij}} \left(\omega_i \underline{j}_j^C - \omega_j \underline{j}_i^C \right) \tag{2.12}$$

with m the average mole mass of the mixture

$$m = \sum_{i=1}^{N} f_i m_i \tag{2.13}$$

Again, in an N component gas mixture there are $N-1$ independent equations of the form of equation 2.12. Together with the additional equation

$$\sum_{i=1}^{N} \underline{j}_i^C = 0 \tag{2.14}$$

they form a closed set of equations from which the N diffusive mass fluxes $\underline{j}_i^C$ can be solved directly. From equation 2.12 we can derive an explicit expression for $\underline{j}_i^C$, assuming that all the other fluxes $\underline{j}_j^C$ are known

$$\underline{j}_i^C = -\rho \mathbb{D}_i \nabla \omega_i - \rho \omega_i \mathbb{D}_i \nabla (\ln m) + m \omega_i \mathbb{D}_i \sum_{\substack{j=1 \\ j \neq i}}^{N} \frac{\underline{j}_j^C}{m_j D_{ij}} \tag{2.15}$$

with $\mathbb{D}_i$ an effective diffusion coefficient for the ith species

$$\mathbb{D}_i = \left(\sum_{\substack{j=1 \\ j \neq i}}^{N} \frac{f_j}{D_{ij}} \right)^{-1} \tag{2.16}$$

The above expressions 2.15 and 2.16 can be used to solve the set of Stefan-Maxwell equations iteratively.

As an alternative to the Stefan-Maxwell equations an approximate expression for the diffusive fluxes in a multicomponent gas mixture can be derived. Here the diffusion of the ith species in a multicomponent mixture is written in the form of Ficks law of diffusion with an effective diffusion coefficient $\mathbb{D}_i'$:

$$\underline{j}_i^C = -\rho \mathbb{D}_i' \nabla \omega_i \tag{2.17}$$

with

$$\mathbb{D}_i' = (1 - f_i) \left(\sum_{\substack{j=1 \\ j \neq i}}^{N} \frac{f_j}{D_{ij}} \right)^{-1} \tag{2.18}$$

In the case of a highly diluted species ($f_i, \omega_i \ll 1$) this approximation is identical with the full Stefan-Maxwell equations. For binary mixtures ($N = 2$) both approaches lead to Ficks law of diffusion. However, when the approximate approach is used to calculate the diffusive fluxes in a multicomponent mixture, the N equations of the form of equation 2.17 are not consistent with equation 2.14. Therefore, in order to be able to fulfill this constraint, one of these equations must be dropped and be replaced by equation 2.14.

2.4.3 Thermal diffusion

Due to the effect of thermal diffusion (or Soret effect), the gas species in an initially homogeneous gas mixture will separate under the influence of a temperature gradient. This effect is usually small compared to ordinary diffusion, but in CVD in coldwall reactors thermal diffusion may be an important effect because of the large temperature gradients present. The thermal diffusion effect cannot easily be explained from simple kinetic considerations based on the concept of mean free path, but is obtained from the rigid kinetic theory developed by Chapman and Enskog. Overviews on thermal diffusion theory can be found in [2.3-2.5]. In general, thermal diffusion causes large, heavy gas molecules to concentrate in cold regions of the reactor, whereas small, light molecules concentrate in the hotter parts of the reactor.

The diffusive mass fluxes due to thermal diffusion are given by

$$\underline{j}_i^T = -\mathbb{D}_i^T \nabla(\ln T) \tag{2.19}$$

in which $\mathbb{D}_i^T$ is the multicomponent thermal diffusion coefficient for the ith species. In general, $\mathbb{D}_i^T$ is a function of the temperature and the composition of the gas mixture, but is independent of the pressure, and $\mathbb{D}_i^T > 0$ for large, heavy molecules and $\mathbb{D}_i^T < 0$ for small, light molecules. Note that $\mathbb{D}_i^T$ is defined in such a way that its dimension equals the dimension of ρD_{ij}, the value of which is also independent of the pressure.

2.4.4　Gas phase reactions

The last term in the concentration equation 2.7 represents the creation and destruction of the ith species due to homogeneous gas phase reactions. Assume that this is a result of K reversible chemical reactions. Since different species may act as products and reactants in each of the K reactions, we use the following general notation for these reactions:

$$\sum_{i=1}^{N} \| - \nu_{ik} \| \mathcal{A}_i \underset{k_{-k}}{\overset{k_k}{\rightleftharpoons}} \sum_{i=1}^{N} \| \nu_{ik} \| \mathcal{A}_i \tag{2.20}$$

Here, the $\mathcal{A}_i$ (i=1,N) represent the different gaseous species, k_k the forward reaction rate constant and k_{-k} the reverse reaction rate constant of the kth gas phase reaction (k=1,K) and ν_{ik} the stoichiometric coefficient for the ith species in the kth gas phase reaction. Note that in the above notation for the reversible reactions, the summation is over the same N species on both sides of the reaction. Now, by taking $\nu_{ik} > 0$ for the products of the forward reaction and $\nu_{ik} < 0$ for the reactants of the forward reaction and by defining $\|\nu_{ik}\| = 0.5(| \nu_{ik} | + \nu_{ik})$ equation 2.20 represents

Table 2.3:   Thermodynamic properties of some gases in W-CVD

gas	H^0_{298} $kJ \cdot mole^{-1}$	S^0_{298} $J \cdot mole^{-1} \cdot K^{-1}$	a_0	$a_1 \cdot 10^3$	$a_2 \cdot 10^6$
Ar	0	154.8	520	0	0
GeF_2	-573	260	434^1	117^1	-50^1
GeH_4	91	217	177	1668	-761
H_2	0	130.7	14622	-1215	1760
He	0	126.1	5192	0	0
HF	-271.1	173.8	1473	-103	137
N_2	0	191.6	1026	5	134
SiF_4	-1615	282.5	378	1265	-657
$SiHF_3$	-1200.8	277.3	333	1566	-764
SiH_4	34.3	204.0	470	3313	-1148
WF_6	-1722	341.1	231	639	-355

$c_p = a_0 + a_1 T + a_2 T^2 \; J \cdot kg^{-1} \cdot K^{-1} \; (300 \; K \leq \mathrm{T} \leq 1000 \; K)$

[1] estimated as: $(c_p m)_{GeF_2} = \frac{(c_p m)_{CF_2}}{(c_p m)_{CF_4}} \cdot (c_p m)_{GeF_4}$

a general equilibrium reaction, with the reactants appearing on the left-hand side and the products appearing on the right-hand side.

When the forward reaction rate constant k_k and the reverse reaction rate constant k_{-k} are known, the reaction rates $\mathcal{R}^g_k$ and $\mathcal{R}^g_{-k}$ may be obtained from

$$\mathcal{R}^g_k = k_k \cdot \prod_{i=1}^{N} c_i^{\|-\nu_{ik}\|} = k_k \cdot \prod_{i=1}^{N} \left(f_i \frac{P}{RT} \right)^{\|-\nu_{ik}\|} \tag{2.21}$$

and

$$\mathcal{R}^g_{-k} = k_{-k} \cdot \prod_{i=1}^{N} c_i^{\|\nu_{ik}\|} = k_{-k} \cdot \prod_{i=1}^{N} \left(f_i \frac{P}{RT} \right)^{\|\nu_{ik}\|} \tag{2.22}$$

with c_i $(=P f_i / RT)$ the mole concentration of the ith species, f_i the mole fraction of the ith species and P the total pressure. Usually, the values of k_k and k_{-k} depend strongly on the temperature and are independent of the pressure for sufficiently high pressures. At low pressures however, k_k and k_{-k} may enter the so-called "pressure fall-off regime". So, in general

$$k_k = k_k(P, T) \tag{2.23}$$

$$k_{-k} = k_{-k}(P, T) \tag{2.24}$$

When either k_k or k_{-k} is known, the reaction rate constant in the opposite direction may be calculated from the reaction equilibrium thermochemistry. Using tabulated values for the standard heat of formation $H^0_{298,i}$ and standard entropy $S^0_{298,i}$ at $P^0 = 1\ atm$ and $T^0 = 298.15\ K$ and values for the specific heat $c_{p,i}$ as a function of temperature [2.6-2.8], the standard Gibbs energy change of formation $G^0_i(T)$ at temperature T for the ith species may be calculated from

$$G^0_i(T) = H^0_{298,i} - TS^0_{298,i} + m_i \int_{T^0}^{T} c_{p,i}(T)dT - Tm_i \int_{T^0}^{T} \frac{c_{p,i}(T)}{T}dT \tag{2.25}$$

Values for the thermodynamic properties H^0_{298}, S^0_{298} and c_p for several common gases in tungsten CVD processes are given in table 2.3. Now we define the standard Gibbs energy change ΔG^0_k for the kth reaction as

$$\Delta G^0_k(T) = \sum_{i=1}^{N} \nu_{ik} G^0_i(T) \tag{2.26}$$

The equilibrium constant K_k for this reaction, defined as

$$K_k = \prod_{i=1}^{N} \left(f_{i,eq} \frac{P}{P^0} \right)^{\nu_{ik}} \tag{2.27}$$

with $f_{i,eq}$ the mole fraction of the ith species in equilibrium, may then be calculated from

$$K_k(T) = e^{-\frac{\Delta G^0_k(T)}{RT}} \tag{2.28}$$

¿From *eqs.* 2.21, 2.22 and 2.27 it may be deduced, that $k_k(P,T)$ and $k_{-k}(P,T)$ are related through

$$k_{-k}(P, T) = \frac{k_k(P, T)}{K_k(T)} \left(\frac{RT}{P^0} \right)^{\sum_{i=1}^{N} \nu_{ik}} \tag{2.29}$$

2.4.5   Final equations for species concentrations

We may now combine all the previous expressions for species transport and chemical reactions in CVD gas mixtures. Thus, combining the general equations (2.7, 2.8) for species transport and chemical reactions, the approximate expressions for ordinary diffusion (2.17, 2.18), the expression for thermal diffusion (2.19) and the expressions for the gas phase reactions (2.21, 2.22) we find the following approximate equation for the species concentrations:

$$\frac{\partial \rho \omega_i}{\partial t} = -\nabla \cdot (\rho \underline{v} \omega_i) + \nabla \cdot (\rho \mathbb{D}_i' \nabla \omega_i) + \nabla \cdot (\mathbb{D}_i^T \nabla (\ln T))$$

$$+ m_i \sum_{k=1}^{K} \nu_{ik} \left(k_k \prod_{i=1}^{N} c_i^{\|-\nu_{ik}\|} - k_{-k} \prod_{i=1}^{N} c_i^{\|\nu_{ik}\|} \right) \tag{2.30}$$

When using the exact Stefan-Maxwell equations (2.15, 2.16) instead of the approximate equations (2.17, 2.18) we have

$$\frac{\partial \rho \omega_i}{\partial t} = -\nabla \cdot (\rho \underline{v} \omega_i) + \nabla \cdot (\rho \mathbb{D}_i \nabla \omega_i) + \nabla \cdot (\rho \omega_i \mathbb{D}_i \nabla (\ln m))$$

$$- \nabla \cdot \left(m \omega_i \mathbb{D}_i \sum_{\substack{j=1 \\ j \neq i}}^{N} \frac{\underline{j}_j^c}{m_j D_{ij}} \right) + \nabla \cdot (\mathbb{D}_i^T \nabla (\ln T))$$

$$+ m_i \sum_{k=1}^{K} \nu_{ik} \left(k_k \prod_{i=1}^{N} c_i^{\|-\nu_{ik}\|} - k_{-k} \prod_{i=1}^{N} c_i^{\|\nu_{ik}\|} \right) \tag{2.31}$$

In 2D cylindrical coordinates, these equations can be written as:

$$\frac{\partial}{\partial t}(r\rho\omega_i) = -\frac{\partial}{\partial r}(r\rho v_r \omega_i) - \frac{\partial}{\partial z}(r\rho v_z \omega_i)$$

$$+ \frac{\partial}{\partial r}(r\rho \mathbb{D}_i' \frac{\partial \omega_i}{\partial r}) + \frac{\partial}{\partial z}(r\rho \mathbb{D}_i' \frac{\partial \omega_i}{\partial z})$$

$$+ \frac{\partial}{\partial r}\left(r\frac{\mathbb{D}_i^T}{T}\frac{\partial T}{\partial r} \right) + \frac{\partial}{\partial z}\left(r\frac{\mathbb{D}_i^T}{T}\frac{\partial T}{\partial z} \right)$$

$$+ r m_i \sum_{k=1}^{K} \nu_{ik} \left(k_k \prod_{i=1}^{N} c_i^{\|-\nu_{ik}\|} - k_{-k} \prod_{i=1}^{N} c_i^{\|\nu_{ik}\|} \right)$$

$$\frac{\partial}{\partial t}(r\rho\omega_i) = -\frac{\partial}{\partial r}(r\rho v_r \omega_i) - \frac{\partial}{\partial z}(r\rho v_z \omega_i)$$

$$+ \frac{\partial}{\partial r}(r\rho \mathbb{D}_i \frac{\partial \omega_i}{\partial r}) + \frac{\partial}{\partial z}(r\rho \mathbb{D}_i \frac{\partial \omega_i}{\partial z})$$

$$+ \frac{\partial}{\partial r}(\frac{r\rho\omega_i \mathbb{D}_i}{m}\frac{\partial m}{\partial r}) + \frac{\partial}{\partial z}(\frac{r\rho\omega_i \mathbb{D}_i}{m}\frac{\partial m}{\partial z})$$

$$- \frac{\partial}{\partial r}\left(r m \omega_i \mathbb{D}_i \sum_{\substack{j=1 \\ j \neq i}}^{N} \frac{\underline{j}_{j,r}^c}{m_j D_{ij}} \right) - \frac{\partial}{\partial z}\left(r m \omega_i \mathbb{D}_i \sum_{\substack{j=1 \\ j \neq i}}^{N} \frac{\underline{j}_{j,z}^c}{m_j D_{ij}} \right)$$

$$+ \frac{\partial}{\partial r}\left(r\frac{\mathbb{D}_i^T}{T}\frac{\partial T}{\partial r} \right) + \frac{\partial}{\partial z}\left(r\frac{\mathbb{D}_i^T}{T}\frac{\partial T}{\partial z} \right)$$

$$+rm_i \sum_{k=1}^{K} \nu_{ik} \left(k_k \prod_{i=1}^{N} c_i^{\|-\nu_{ik}\|} - k_{-k} \prod_{i=1}^{N} c_i^{\|\nu_{ik}\|} \right)$$

2.5   Surface reactions

At the wafer surface, a total number of L surface reactions, transforming gaseous reactants into solid and gaseous reaction products, will take place of the form

$$\sum_{i=1}^{N} \|-\sigma_{il}\| \mathcal{A}_i + \sum_{j=1}^{M} \|-\chi_{jl}\| \mathcal{B}_j \xrightarrow{\mathcal{R}_l^s} \sum_{i=1}^{N} \|\sigma_{il}\| \mathcal{A}_i + \sum_{j=1}^{M} \|\chi_{jl}\| \mathcal{B}_j \ (l = 1, L)$$

$$(2.32)$$

with $\mathcal{A}_i$ the gaseous reactants and reaction products ($i=1,N$) and $\mathcal{B}_j$ the solid reactants and reaction products ($j=1,M$). The σ_{il} and χ_{jl} represent the stoichiometric coefficients for the ith gaseous and the jth solid species in the lth surface reaction ($l=1,L$). With respect to the sign of the stoichiometric coefficients the same convention has been followed as in section 2.4.4. From *eq.* 2.32 the growth rate $\mathcal{G}_j$ of the jth solid species in $\mathring{A}ng/min$ may be deduced:

$$\mathcal{G}_j = 60 \cdot 10^{10} \frac{m_j}{\rho_j^s} \sum_{l=1}^{L} \mathcal{R}_l^s \chi_{jl} \tag{2.33}$$

with m_j the mole mass of species $\mathcal{B}_j$ in $kg \cdot mole^{-1}$, ρ_j^s its density in the solid phase in $kg \cdot m^{-3}$ and $\mathcal{R}_l^s$ the surface reaction rate in $mole \cdot m^{-2} \cdot s^{-1}$.

Usually, heterogeneous surface reactions are characterized by complicated mechanisms consisting of a number of different reaction steps. Amongst others, the surface reaction speed $\mathcal{R}_l^s$ will depend in a complicated way on the partial pressures of the gaseous species, the rate constants of the individual reaction steps, and the surface coverage. However, usually little or no information on the individual reaction steps and rate constants is available. Therefore, what is often done is to propose a mechanism and to assume that one of the steps in this mechanism is rate limiting whereas all the other steps are in equilibrium. With these assumptions an expression for the rate of the overall reaction in the form

$$\mathcal{R}^s = \mathcal{R}^s(P_1, ..., P_N, T_S) \tag{2.34}$$

may be found, with P_i the partial pressure of the ith gaseous species and T_S the surface temperature. From a comparison of predictions from this expression with the experimentally observed growth rates one can now judge the validity of the assumptions, and make an estimate of the unknown constants in the expression 2.34.

A concept often used to give expressions for the surface reaction rate of a mono- molecular surface dissociation reaction $A(g) \longrightarrow B(s)+C(g)$ is the so called "reactive sticking coefficient" [e.g. 2.9-2.13]. It is defined as the ratio of the number of molecules B which are incorporated into the deposited film and the number of collisions of molecules A with the surface per unit time. From kinetic theory it can be derived that the mole flux $\mathcal{F}_A$ of gaseous molecules A colliding with the surface equals

$$\mathcal{F}_A = \frac{P_A}{(2\pi m_A R T_S)^{\frac{1}{2}}} \tag{2.35}$$

Defining a reactive sticking coefficient γ_A for species A we now find

$$\mathcal{R}^s = \gamma_A \frac{P_A}{(2\pi m_A R T_S)^{\frac{1}{2}}} \tag{2.36}$$

with $\gamma_A \leq 1$. In general, γ_A is not a constant, but depends on the surface temperature and surface coverage with adsorbed species.

2.6 Boundary conditions

For every transport equation described in sections 2.2-2.4 a set of boundary conditions must be prescribed. Thus, boundary conditions are needed for the velocity, the temperature and the species concentrations on nonreacting walls, on the reacting surfaces and in the inflow and outflow of the reactor. The pressure must be prescribed in one point. In the following these different boundary conditions are discussed.

Nonreacting walls
On nonreacting walls the no-slip and impermeability conditions apply for the velocity

$$\underline{v} = \underline{0} \tag{2.37}$$

Prescribed temperature boundary conditions and zero temperature gradients normal to the wall apply for isothermal and adiabatic walls, respectively

$$
\begin{aligned}
T &= T_{wall} \ (isothermal \ walls) \tag{2.38} \\
\underline{n} \cdot \nabla T &= 0 \ (adiabatic \ walls) \tag{2.39}
\end{aligned}
$$

The total mass flux vector normal to a nonreacting surface must be zero for each of the species:

$$\underline{n} \cdot (\underline{j}_i^C + \underline{j}_i^T) = 0 \qquad (2.40)$$

where $\underline{n}$ is a unity vector normal to the surface of the wall.

Reacting surfaces

Due to the surface reactions (*eq.* 2.32) there will be a net mass production rate $\mathcal{P}_i$ of gaseous the ith species at the wafer surface

$$\mathcal{P}_i = m_i \sum_{l=1}^{L} \sigma_{il} \mathcal{R}_l^s \qquad (2.41)$$

Thus, the velocity component normal to the wafer surface is expressed by

$$\underline{n} \cdot \underline{v} = \frac{1}{\rho} \sum_{i=1}^{N} m_i \sum_{l=1}^{L} \sigma_{il} \mathcal{R}_l^s \qquad (2.42)$$

with $\underline{n}$ a unity vector normal to the reacting surface. Assuming the no-slip condition to hold, the tangential component is zero:

$$\underline{n} \times \underline{v} = \underline{0} \qquad (2.43)$$

Usually, the reacting surfaces will be more or less isothermal, leading to

$$T = T_{wall} \qquad (2.44)$$

The net total mass flux of the ith species normal to the wafer surface must be equal to $\mathcal{P}_i$, so

$$\underline{n} \cdot (\rho \omega_i \underline{v} + \underline{j}_i^C + \underline{j}_i^T) = m_i \sum_{l=1}^{L} \sigma_{il} \mathcal{R}_l^s \qquad (2.45)$$

Inflow

If an inflow of Q_i standard liters per minute (*slm*, *i.e.* liters per minute at standard temperature T^0 and pressure P^0) is prescribed in the inflow of the reactor for each species, the inflow velocity follows from

$$v_{in} = \frac{10^{-3}}{60} \frac{P^0}{T^0} \frac{T_{in}}{P_{in}} \frac{1}{A_{in}} \sum_{i=1}^{N} Q_i \qquad (2.46)$$

with v_{in} the inflow velocity in m/s, A_{in} the surface area of the inflow opening in m^2, P_{in} the pressure at the inlet of the reactor and T_{in} the inflow temperature. Then, the boundary conditions for the velocity vector in the inflow are given by

$$\underline{n} \cdot \underline{v} = v_{in} \tag{2.47}$$

$$\underline{n} \times \underline{v} = \underline{0} \tag{2.48}$$

with $\underline{n}$ a unity vector normal to the inflow opening.
The temperature in the inflow is prescribed as

$$T = T_{in} \tag{2.49}$$

The total mass flow of the ith species into the reactor must correspond to Q_i according to

$$\underline{n} \cdot (\rho \omega_i \underline{v} + \underline{j}_i^C + \underline{j}_i^T) = \frac{10^{-3}}{60} \frac{P^0}{T^0} \frac{Q_i m_i}{R A_{in}} \tag{2.50}$$

with the fluxes in $kg \cdot m^{-2} \cdot s^{-1}$, Q in slm, A in m^2, P^0 in Pa, T^0 in K, m in $kg/mole$ and the gas constant R in $J \cdot mole^{-1} \cdot K^{-1}$. This is most easily done by taking each of the inlet concentrations ω_i fixed as

$$\omega_i = \frac{m_i Q_i}{\sum\limits_{j=1}^{N} m_j Q_j} \tag{2.51}$$

and by prohibiting species diffusion through the inflow

$$\underline{n} \cdot (\underline{j}_i^C + \underline{j}_i^T) = 0 \tag{2.52}$$

Outflow
In the outflow zero gradients for the total mass flux vector in the direction normal to the outflow opening may be assumed, as well as zero heat and species diffusion fluxes. Furthermore, we assume that the direction of the velocity is normal to the outflow opening

$$\underline{n} \cdot (\nabla \rho \underline{v}) = 0, \quad \underline{n} \times \underline{v} = \underline{0} \tag{2.53}$$

$$\underline{n} \cdot (\lambda \nabla T) = 0 \tag{2.54}$$

$$\underline{n} \cdot (\underline{j}_i^C + \underline{j}_i^T) = 0 \tag{2.55}$$

with $\underline{n}$ a unity vector normal to the outflow opening.

Modeling of wall temperatures
It is not always possible to use simple boundary conditions such as the isothermal boundary condition (*eq.* 2.38) or the adiabatic boundary condition (*eq.* 2.39) for the temperatures of solid walls. In CVD reactors

we often have to do with reactor walls and wafers which adopt a certain temperature profile as a result of the conjugate heat exchange with their surroundings. In that case, side wall temperatures may be obtained from detailed thermal energy balance models, accounting for heat conduction in the wall material, and conductive, radiative and convective heat exchange between susceptor, reactor gases, reactor walls and surroundings. An example of such modeling will be discussed in chapter 3.

2.7 Prediction of transport properties

In the absence of experimental data, the transport properties of gas species and gas mixtures may be calculated from kinetic theory. This is discussed in the following sections.

2.7.1 Properties of gas species

The thermal conductivity and the dynamic viscosity of a gas species are functions of the temperature and, for the pressures common in CVD, independent of the pressure. For gases like H_2, He, N_2 and Ar experimental data are available [2.14-2.16]. For less common gases, the thermal conductivity and dynamic viscosity may be predicted from kinetic theory [2.5, 2.17-2.19]. In doing so, assumptions have to be made on the form of the intermolecular potential energy function $\phi(r)$. For non-polar molecules a commonly used and reasonably accurate intermolecular potential energy function is the Lennard-Jones potential:

$$\phi(r) = 4\epsilon \left[(\frac{\sigma}{r})^{12} - (\frac{\sigma}{r})^6 \right] \qquad (2.56)$$

where r is the distance between the molecules, σ the collision diameter of the molecules and ϵ the maximum energy of attraction. Using the Lennard-Jones potential, the ith gas species is characterized by three parameters: its mole mass m_i (in $kg/mole$), its collision diameter σ_i (in $\mathring{A}ng$) and the maximum energy of attraction ϵ_i, usually given as $(\epsilon/k)_i$ (in K, with k Boltzmanns constant). These parameters may be obtained from experimental viscosity or thermal conductivity data. They may also be estimated from properties of the gas at the critical point or at the boiling point [2.2, p. 22]:

$$\frac{\epsilon}{k} \cong 0.77T_c \text{ or } \frac{\epsilon}{k} \cong 1.15T_b \qquad (2.57)$$

Table 2.4:   Lennard-Jones parameters of some
common gases in W-CVD

gas	m (kg/mole)	σ (Ång)	ϵ/k (K)
Ar	0.03994	3.542	93.3
GeF_2	0.11059	4.66^1	164^1
GeH_4	0.07662	4.40^2	225^2
H_2	0.00202	2.827	59.7
He	0.00400	2.58	10.2
HF	0.02001	3.138	330
N_2	0.02802	3.798	71.4
SiF_4	0.10409	4.880	171.9
$SiHF_3$	0.08609	4.53^3	308^3
SiH_4	0.03212	4.084	207.6
WF_6	0.29784	5.21^2	338^2

[1] estimated as: $\sigma(GeF_2) = \frac{\sigma(CF_2)}{\sigma(CF_4)} \cdot \sigma(GeF_4)$
and similar for ϵ/k

[2] estimated from fluid properties at
the critical and boiling point

[3] estimated as: $\sigma(SiHF_3) = \frac{\sigma(CHF_3)}{\sigma(CF_4)} \cdot \sigma(SiF_4)$
and similar for ϵ/k

$$\sigma \cong 0.841 V_c^{1/3} \text{ or } \sigma \cong 2.44(\frac{T_c}{P_c})^{1/3} \text{ or } \sigma \cong 1.166 V_{b,l}^{1/3} \qquad (2.58)$$

where T_c and T_b are the critical temperature and normal boiling point
temperature in K, P_c is the critical pressure in atmospheres, V_c is the
molar volume at the critical point in $cm^3/mole$ and $V_{b,l}$ the molar volume
of the liquid at the normal boiling point in $cm^3/mole$. Values for m, σ
and ϵ/k for some common gases in W-CVD are given in table 2.4.
A reduced temperature T^* is now defined as $T^* = T \cdot (\epsilon/k)^{-1}$ and an
integral function $\Omega_\mu(T^*)$ is introduced. This slowly varying function of T^*
has been tabulated in [2.5, pp. 1126-1127]. Polynomials for calculating
Ω_μ are given in table 2.5. For the dynamic viscosity μ_i of the ith species

Table 2.5: Polynomials for calculating Ω_μ, Ω_D, A^*, B^* and C^*

$$\phi = a_0 + a_1 T^* + a_2 T^{*2} + a_3 T^{*3}$$

	a_0	a_1	a_2	a_3
Ω_μ:				
$0.3 \leq T^* < 1$	$+4.0384 \cdot 10^0$	$-5.2953 \cdot 10^0$	$+4.0846 \cdot 10^0$	$-1.2414 \cdot 10^0$
$1 \ \leq T^* < 3$	$+2.7015 \cdot 10^0$	$-1.6152 \cdot 10^0$	$+5.6831 \cdot 10^{-1}$	$-7.1633 \cdot 10^{-2}$
$3 \ \leq T^* < 10$	$+1.3824 \cdot 10^0$	$-1.6174 \cdot 10^{-1}$	$+1.7603 \cdot 10^{-2}$	$-7.1058 \cdot 10^{-4}$
$10 \leq T^* < 30$	$+9.8307 \cdot 10^{-1}$	$-1.9520 \cdot 10^{-2}$	$+4.4911 \cdot 10^{-4}$	$-3.8432 \cdot 10^{-6}$
Ω_D:				
$0.3 \leq T^* < 1$	$+4.2639 \cdot 10^0$	$-7.1400 \cdot 10^0$	$+6.6722 \cdot 10^0$	$-2.3599 \cdot 10^0$
$1 \ \leq T^* < 3$	$+2.4130 \cdot 10^0$	$-1.4086 \cdot 10^0$	$+4.9307 \cdot 10^{-1}$	$-6.2149 \cdot 10^{-2}$
$3 \ \leq T^* < 10$	$+1.2729 \cdot 10^0$	$-1.5195 \cdot 10^{-1}$	$+1.6376 \cdot 10^{-2}$	$-6.4903 \cdot 10^{-4}$
$10 \leq T^* < 30$	$+8.9722 \cdot 10^{-1}$	$-1.9064 \cdot 10^{-2}$	$+4.4207 \cdot 10^{-4}$	$-3.7962 \cdot 10^{-6}$
A^*:				
$0.3 \leq T^* < 1$	$+8.7140 \cdot 10^{-1}$	$+8.5547 \cdot 10^{-1}$	$+1.0310 \cdot 10^0$	$+4.0751 \cdot 10^{-1}$
$1 \ \leq T^* < 5$	$+1.1253 \cdot 10^0$	$-3.1927 \cdot 10^{-2}$	$+9.6617 \cdot 10^{-3}$	$-8.7623 \cdot 10^{-4}$
$5 \ \leq T^* < 30$	$+1.0929 \cdot 10^0$	$+2.1304 \cdot 10^{-3}$	$-4.9384 \cdot 10^{-5}$	$+4.3252 \cdot 10^{-7}$
B^*:				
$0.3 \leq T^* < 1$	$+1.1719 \cdot 10^0$	$+7.5414 \cdot 10^{-1}$	$-1.3929 \cdot 10^0$	$+6.6028 \cdot 10^{-1}$
$1 \ \leq T^* < 3$	$+1.3878 \cdot 10^0$	$-2.8081 \cdot 10^{-1}$	$+9.5936 \cdot 10^{-2}$	$-1.1401 \cdot 10^{-2}$
$3 \ \leq T^* < 10$	$+1.1255 \cdot 10^0$	$-1.2018 \cdot 10^{-2}$	$+1.2583 \cdot 10^{-3}$	$-3.6681 \cdot 10^{-5}$
$10 \leq T^* < 30$	$+1.0950 \cdot 10^0$	$+0.0000 \cdot 10^0$	$+0.0000 \cdot 10^0$	$+0.0000 \cdot 10^0$
C^*:				
$0.3 \leq T^* < 1$	$+9.4008 \cdot 10^{-1}$	$-4.5832 \cdot 10^{-1}$	$+5.5896 \cdot 10^{-1}$	$-2.0409 \cdot 10^{-1}$
$1 \ \leq T^* < 3$	$+7.6201 \cdot 10^{-1}$	$+9.0169 \cdot 10^{-2}$	$-1.6830 \cdot 10^{-2}$	$+1.0998 \cdot 10^{-3}$
$3 \ \leq T^* < 10$	$+8.3975 \cdot 10^{-1}$	$+3.4106 \cdot 10^{-2}$	$-3.8932 \cdot 10^{-3}$	$+1.5349 \cdot 10^{-4}$
$10 \leq T^* < 30$	$+9.4335 \cdot 10^{-1}$	$+1.6500 \cdot 10^{-4}$	$+0.0000 \cdot 10^0$	$+0.0000 \cdot 10^0$

we now obtain from kinetic theory [2.5, p. 527]

$$\mu_i = \frac{5}{16} \frac{(\pi m_i RT)^{\frac{1}{2}}}{\pi \sigma_i^2 N_A \Omega_\mu(T_i^*)} \tag{2.59}$$

with m_i the mole mass and σ_i the collision diameter of the ith species, R the universal gas constant and N_A Avogadro's number. Using SI units this expression can be written as

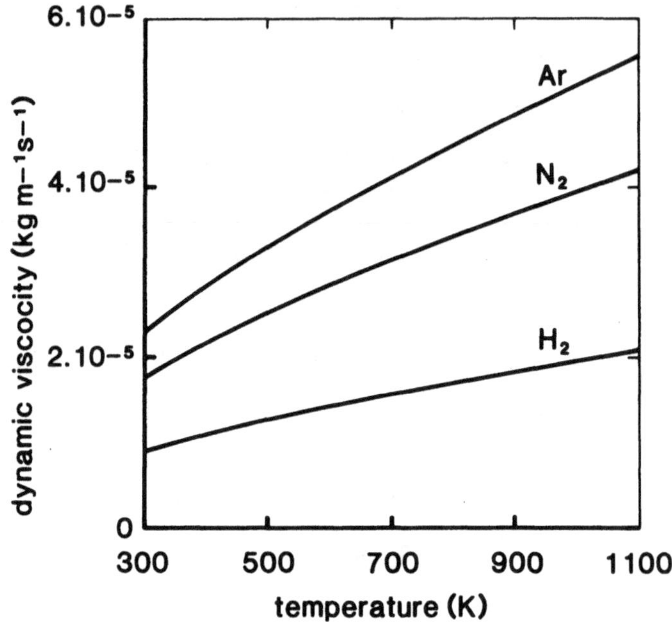

Figure 2.1: Viscosity of some common CVD carrier gases

$$\mu_i = 8.441 \cdot 10^{-5} \frac{(m_i T)^{\frac{1}{2}}}{\sigma_i^2 \Omega_\mu(T_i^*)} \qquad (2.60)$$

with m_i in $kg \cdot mole^{-1}$, σ_i in $\AA ng$, T in K and μ_i in $kg \cdot m^{-1} \cdot s^{-1}$. The thermal conductivity λ_i of a mono-atomic gas species i predicted by kinetic theory is [2.5, p. 534]

$$\lambda_i = \frac{15}{4} \frac{R}{m_i} \mu_i \qquad (2.61)$$

For poly-atomic gases, several semi-empirical corrections to the relations predicted by kinetic theory have been proposed in order to account for the transfer of energy between internal degrees of freedom and translational motion [2.18]. A reasonably accurate expression is given by the so-called modified Eucken correction [2.19]

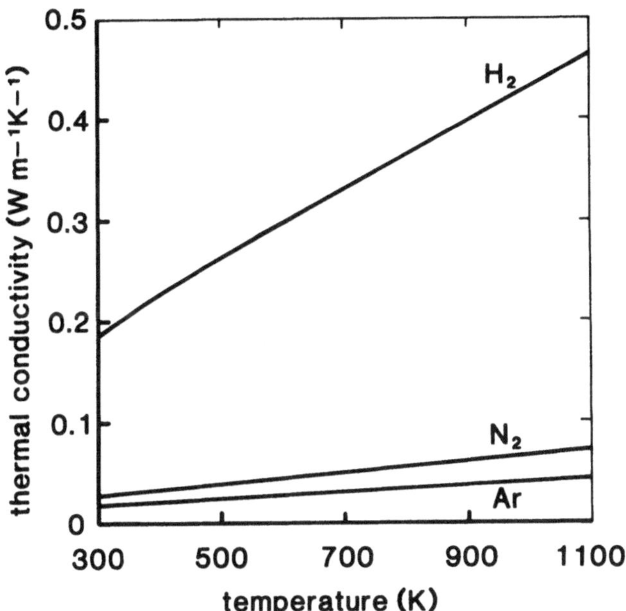

Figure 2.2: Thermal conductivity of some common CVD carrier gases

$$\lambda_i = \left[\frac{15}{4} + 1.32\left(\frac{c_{p,i}m_i}{R} - \frac{5}{2}\right)\right]\frac{R}{m_i}\mu_i \qquad (2.62)$$

with m_i in $kg \cdot mole^{-1}$, $c_{p,i}$ in $J \cdot kg^{-1} \cdot K^{-1}$, μ_i in $kg \cdot m^{-1} \cdot s^{-1}$, R in $J \cdot mole^{-1} \cdot K^{-1}$ and λ_i in $W \cdot m^{-1} \cdot K^{-1}$. For a mono-atomic gas with $c_{p,i}m_i = 5R/2$, *eqs.* 2.61 and 2.62 are identical.

Figures 2.1 and 2.2 show viscosities and thermal conductivities of some common CVD carrier gases as a function of temperature. For the temperature range of interest for CVD applications, the viscosity of most common gases varies with temperature as $T^{0.65-0.70}$ and the thermal conductivity as $T^{0.70-0.75}$.

The bulk viscosity κ is identically zero for low-density, mono-atomic gases and is probably not too important for poly-atomic and dense gases [2.2, p. 79]. Its value is most often set to zero for low-density gas flows.

2.7.2 Properties of gas mixtures

Using the properties of the constituent species, the properties of a gas mixture may be calculated. The density of a gas mixture may of course be calculated straightforwardly from

$$\rho = \sum_{i=1}^{N} \rho_i = \frac{Pm}{RT} \tag{2.63}$$

with m the average mole mass according to *eq. 2.13*.
Several (semi)empirical relations for the viscosity and thermal conductivity of gas mixtures have been evaluated by Bretsznajder [2.17]. Following his recommendations, the following relations can be used:

$$\mu = \sum_{i=1}^{N} \left(\frac{f_i \mu_i}{\sum\limits_{j=1}^{N} f_j \Phi_{ij}} \right) \tag{2.64}$$

with

$$\Phi_{ij} = \frac{1}{\sqrt{8}} \left(1 + \frac{m_i}{m_j} \right)^{-\frac{1}{2}} \cdot \left[1 + \left(\frac{\mu_i}{\mu_j} \right)^{1/2} \cdot \left(\frac{m_j}{m_i} \right)^{1/4} \right]^2 \tag{2.65}$$

and

$$\lambda = \zeta \sum_{i=1}^{N} \lambda_i f_i + (1 - \zeta) \cdot \left(\sum_{i=1}^{N} \frac{f_i}{\lambda_i} \right)^{-1} \tag{2.66}$$

in which ζ is a fraction, the value of which depends on the mole fraction f_L of light gases (He, H_2) and may be approximated by

$$\zeta = +0.312 + 0.325 f_L - 0.311 f_L^2 + 0.469 f_L^3 \tag{2.67}$$

The specific heat of the gas mixture is simply the mass averaged specific heat

$$c_p = \sum_{i=1}^{N} \omega_i c_{p,i} \tag{2.68}$$

2.7.3 Ordinary diffusion coefficients

The binary diffusion coefficient D_{ij} for a gas pair i and j may also be calculated from kinetic theory. Its value depends on the temperature and

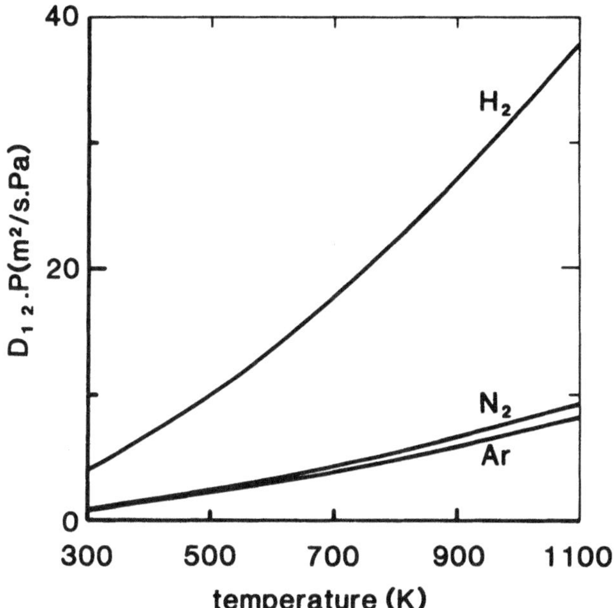

Figure 2.3: Diffusion coefficient of WF_6 in some common CVD carrier gases

the pressure, but is virtually independent of the mixture composition. Lennard-Jones parameters for a gas pair $i - j$ are introduced according to

$$\sigma_{ij} = 0.5(\sigma_i + \sigma_j) \tag{2.69}$$

$$\left(\frac{\epsilon}{k}\right)_{ij} = \left[\left(\frac{\epsilon}{k}\right)_i \left(\frac{\epsilon}{k}\right)_j\right]^{1/2} \tag{2.70}$$

$$T_{ij}^* = \frac{T}{(\epsilon/k)_{ij}} \tag{2.71}$$

The integral function $\Omega_D(T_{ij}^*)$ is a slowly varying function of T_{ij}^* and has been tabulated [2.5, pp.1126-1127]. Polynomials for calculating Ω_D are given in table 2.5. Now, from kinetical theory, we have for the binary diffusion coefficient D_{ij} [2.5, p. 527]

$$D_{ij} = \frac{3}{16}\left(\frac{m_i + m_j}{m_i m_j}\right)^{1/2} \cdot \frac{(2\pi R^3 T^3)^{1/2}}{P N_A \pi \sigma_{ij}^2 \Omega_D(T_{ij}^*)} \tag{2.72}$$

or, in SI units

$$D_{ij} = 5.88 \cdot 10^{-4} \left(\frac{m_i + m_j}{m_i m_j} \right)^{1/2} \cdot \frac{T^{3/2}}{P\sigma_{ij}^2 \Omega_D(T_{ij}^*)} \tag{2.73}$$

with D_{ij} the binary diffusion coefficient in $m^2 \cdot s^{-1}$, m_i and m_j the mole masses in $kg/mole$, P the pressure in Pa and σ the collision diameter in $\mathring{A}ng$. Wilke and Lee [2.20] have shown, that the assumption of a constant numerical coefficient in *eq.* 2.73 is not legitimate and that its value varies with the molecular weight of the diffusing gases. Thus, they suggest

$$D_{ij} = \left[6.77 - 0.0492 \left(\frac{m_i + m_j}{m_i m_j} \right)^{1/2} \right] \cdot 10^{-4} \left(\frac{m_i + m_j}{m_i m_j} \right)^{1/2} \frac{T^{3/2}}{P\sigma_{ij}^2 \Omega_D(T_{ij}^*)} \tag{2.74}$$

Figure 2.3 shows $D_{12}P$ for mixtures of WF_6 and some common CVD carrier gases as a function of temperature. For the temperature range of interest for CVD applications, the binary diffusion coefficient of most gas pairs varies as $T^{1.7-1.8}$.

2.7.4 Thermal diffusion coefficients

In a binary gas mixture, the thermal diffusion coefficient for each of the species is given by:

$$\mathbb{D}_1^T = -\mathbb{D}_2^T = \frac{P}{mRT} m_1 m_2 D_{12} \alpha_{12} f_1 f_2 = \rho D_{12} \alpha_{12} \omega_1 \omega_2 \tag{2.75}$$

where α_{12} is the thermal diffusion factor. When species 1 moves to the cold region, $\alpha_{12} > 0$ and when species 2 moves to the cold region $\alpha_{12} < 0$. For a given gas pair, α_{12} is a function of the temperature and the binary mixture composition, but independent of pressure:

$$\alpha_{12} = -\alpha_{21} = \alpha_{12}(f_1, f_2, T) \tag{2.76}$$

Unlike other transport properties, values of α_{12} predicted from kinetic theory are strongly dependent on the assumed intermolecular potential. Fairly accurate predictions are obtained when the Lennard-Jones potential is assumed. This leads to the following equations for α_{12} [2.5, pp.534-541]:

$$\alpha_{12} = \frac{1}{6\lambda_{12}} \frac{S^{(1)} f_1 - S^{(2)} f_2}{X_\lambda + Y_\lambda} (6C_{12}^* - 5) \tag{2.77}$$

with

$$S^{(1)} = \frac{m_1 + m_2}{2m_2}\frac{\lambda_{12}}{\lambda_1} - \frac{15}{4A_{12}^*}\frac{m_2 - m_1}{2m_1} - 1 \tag{2.78}$$

$$S^{(2)} = \frac{m_1 + m_2}{2m_1}\frac{\lambda_{12}}{\lambda_2} - \frac{15}{4A_{12}^*}\frac{m_1 - m_2}{2m_2} - 1 \tag{2.79}$$

$$\lambda_{12} = 0.00263 \cdot \frac{\left(T\frac{m_1+m_2}{2m_1 m_2}\right)^{1/2}}{\sigma_{12}^2 \Omega_\mu(T_{12}^*)} \tag{2.80}$$

$$\lambda_1 = 0.00263 \cdot \frac{(T/m_1)^{1/2}}{\sigma_1^2 \Omega_\mu(T_1^*)} \quad \text{and similar for } \lambda_2 \tag{2.81}$$

$$X_\lambda = \frac{f_1^2}{\lambda_1} + \frac{2f_1 f_2}{\lambda_{12}} + \frac{f_2^2}{\lambda_2} \tag{2.82}$$

$$Y_\lambda = \frac{f_1^2}{\lambda_1}U^{(1)} + \frac{2f_1 f_2}{\lambda_{12}}U^{(Y)} + \frac{f_2^2}{\lambda_2}U^{(2)} \tag{2.83}$$

$$U^{(1)} = \frac{4}{15}A_{12}^* - \frac{1}{12}\left(\frac{12}{5}B_{12}^* + 1\right)\frac{m_1}{m_2} + \frac{1}{2}\frac{(m_1 - m_2)^2}{m_1 m_2} \tag{2.84}$$

$$U^{(2)} = \frac{4}{15}A_{12}^* - \frac{1}{12}\left(\frac{12}{5}B_{12}^* + 1\right)\frac{m_2}{m_1} + \frac{1}{2}\frac{(m_2 - m_1)^2}{m_1 m_2} \tag{2.85}$$

$$\begin{aligned}
U^{(Y)} = {} & \frac{4}{15}A_{12}^*\frac{(m_1 + m_2)^2}{4m_1 m_2}\frac{\lambda_{12}^2}{\lambda_1 \lambda_2} - \frac{1}{12}\left(\frac{12}{5}B_{12}^* + 1\right) \\
& - \frac{5}{32A_{12}^*}\left(\frac{12}{5}B_{12}^* - 5\right)\frac{(m_2 - m_1)^2}{m_1 m_2} \tag{2.86}
\end{aligned}$$

where the mole mass m should be taken in $kg/mole$. Figure 2.4 shows the binary thermal diffusion factor α_{12} for the Lennard-Jones potential, for different mole fractions of WF_6 in some common carrier gases. When the molecules are assumed to behave as rigid elastic spheres, α_{12} may be calculated from *eqs.* 2.77-2.86 by setting $A^* = B^* = C^* = \Omega_\mu = 1$. For the special case of a highly diluted heavy species 1 in a lighter carrier gas 2 ($f_1 \ll 1$, $m_1 > m_2$, $\sigma_1 > \sigma_2$), the value of α_{12} may then be approximated by

$$\begin{aligned}
\alpha_{12} = {} & 0.89\left(\frac{m_1 - m_2}{m_1 + m_2}\right) + 0.17\left[\left(\frac{\sigma_1 + \sigma_2}{2\sigma_2}\right)^2 - 1\right] \\
& + 0.74\left(\frac{m_1 - m_2}{m_1 + m_2}\right)\left[\left(\frac{\sigma_1 + \sigma_2}{2\sigma_2}\right)^2 - 1\right] \tag{2.87}
\end{aligned}$$

In general however, the rigid elastic sphere assumption leads to a significant over-estimation of thermal diffusion factors.

Using the binary thermal diffusion factors, an approximate expression for

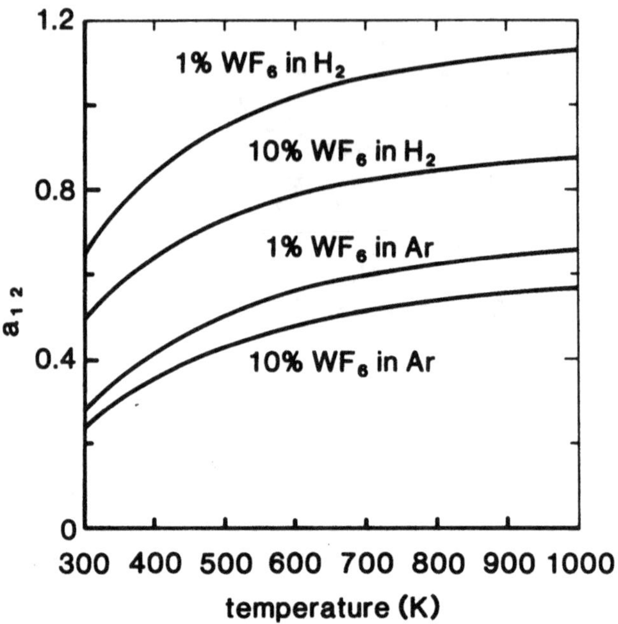

Figure 2.4: Thermal diffusion factor α_{12} for different $WF_6 - H_2$ and $WF_6 - Ar$ mixtures

the multi-component thermal diffusion coefficients, suggested by Clark Jones [2.21], is given by

$$\mathbb{D}_i^T = \sum_{\substack{j=1 \\ j \neq i}}^{N} \frac{P}{mRT} m_i m_j D_{ij} \alpha_{ij} f_i f_j = \sum_{\substack{j=1 \\ j \neq i}}^{N} \rho \omega_i \omega_j D_{ij} \alpha_{ij} \qquad (2.88)$$

which should be compared to *eq.* 2.75. Here, α_{ij} is calculated from *eqs.* 2.77-2.86, replacing f_i by $f_i/(f_i + f_j)$ and f_j by $f_j/(f_i + f_j)$. This approximation is exact for binary mixtures, as well as for multicomponent mixtures of isotopes. Furthermore, this approach ensures that the sum of all thermal diffusion coefficients equals zero, as it should.

The exact calculation of the multicomponent thermal diffusion coefficients in an N component gas mixture involves the computation of the value of one $2N \times 2N$ and N $(2N+1) \times (2N+1)$ determinants, the elements of which are complicated functions of the mole fractions and Lennard-Jones parameters of all species and of the temperature:

$$
\mathbb{D}_k^T = -\frac{8m_k}{5R} \frac{
\begin{vmatrix}
L_{11}^{00} & \cdots & L_{1N}^{00} & L_{11}^{01} & \cdots & L_{1N}^{01} & 0 \\
 & & & & & & \\
\vdots & & \vdots & \vdots & & \vdots & \vdots \\
 & & & & & & \\
L_{N1}^{00} & \cdots & L_{NN}^{00} & L_{N1}^{01} & \cdots & L_{NN}^{01} & 0 \\
L_{11}^{10} & \cdots & L_{1N}^{10} & L_{11}^{11} & \cdots & L_{1N}^{11} & f_1 \\
 & & & & & & \\
\vdots & & \vdots & \vdots & & \vdots & \vdots \\
 & & & & & & \\
L_{N1}^{10} & \cdots & L_{NN}^{10} & L_{N1}^{11} & \cdots & L_{NN}^{11} & f_N \\
f_1\delta_{1k} & \cdots & f_N\delta_{Nk} & 0 & \cdots & 0 & 0
\end{vmatrix}
}{
\begin{vmatrix}
L_{11}^{00} & \cdots & L_{1N}^{00} & L_{11}^{01} & \cdots & L_{1N}^{01} \\
 & & & & & \\
\vdots & & \vdots & \vdots & & \vdots \\
 & & & & & \\
L_{N1}^{00} & \cdots & L_{NN}^{00} & L_{N1}^{01} & \cdots & L_{NN}^{01} \\
L_{11}^{10} & \cdots & L_{1N}^{10} & L_{11}^{11} & \cdots & L_{1N}^{11} \\
 & & & & & \\
\vdots & & \vdots & \vdots & & \vdots \\
 & & & & & \\
L_{N1}^{10} & \cdots & L_{NN}^{10} & L_{N1}^{11} & \cdots & L_{NN}^{11}
\end{vmatrix}
} \tag{2.89}
$$

with $\mathbb{D}_k^T$ the multicomponent thermal diffusion coefficient of the kth species in a N component gas mixture, m_k its mole mass in $kg/mole$, R the universal gas constant in $J/moleK$, δ_{ij} the Kronecker delta function and

$$L_{ii}^{00} = 0 \tag{2.90}$$

$$L_{ij}^{00} = \frac{2f_if_j}{A_{ij}^*\lambda_{ij}} + \sum_{\substack{n=1\\n\neq i}}^{N} \frac{2f_if_nm_j}{m_iA_{in}^*\lambda_{in}} \quad (i \neq j) \tag{2.91}$$

$$L_{ii}^{01} = 5\sum_{\substack{n=1\\n\neq i}}^{N} \frac{f_if_nm_n(\frac{6}{5}C_{in}^* - 1)}{(m_i+m_n)A_{in}^*\lambda_{in}} \tag{2.92}$$

$$L_{ij}^{01} = -5\frac{f_if_jm_i(\frac{6}{5}C_{ij}^* - 1)}{(m_i+m_j)A_{ij}^*\lambda_{ij}} \quad (i \neq j) \tag{2.93}$$

$$L_{ij}^{10} = \frac{m_j}{m_i}L_{ij}^{01} \tag{2.94}$$

$$
\begin{aligned}
L_{ii}^{11} = &-\frac{4f_i^2}{\lambda_i} \\
&- \sum_{\substack{n=1\\n\neq i}}^{N} \frac{2f_if_n\left(\frac{15}{2}m_i^2 + \frac{25}{4}m_n^2 - 3m_n^2B_{in}^* + 4m_im_nA_{in}^*\right)}{(m_i+m_n)^2A_{in}^*\lambda_{in}}
\end{aligned} \tag{2.95}
$$

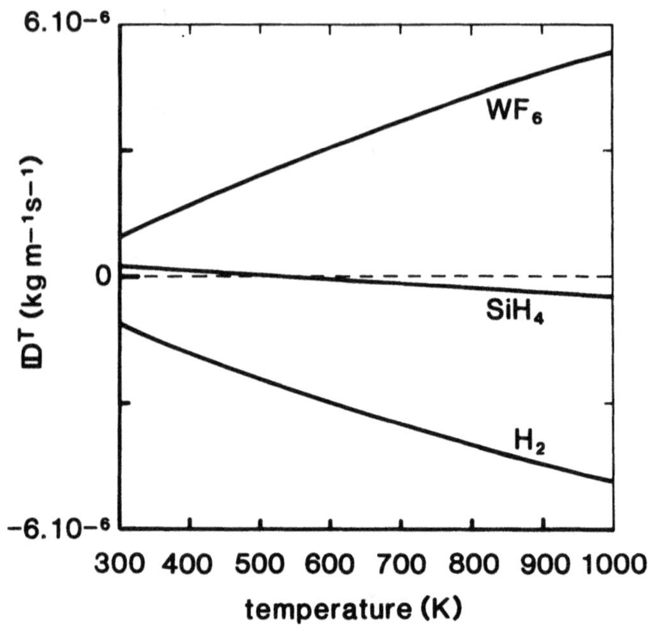

Figure 2.5: Multicomponent thermal diffusion coefficients in a 80 mole % H_2 + 10 mole % WF_6 + 10 mole % SiH_4 mixture

$$L_{ij}^{11} = \frac{2f_i f_j m_i m_j}{(m_i + m_j)^2 A_{ij}^* \lambda_{ij}} \left(\frac{55}{4} - 3B_{ij}^* - 4A_{ij}^* \right) \ (i \neq j) \qquad (2.96)$$

with λ in $W \cdot m^{-1} \cdot K^{-1}$. This approach demands a great computational effort. However, especially in gas mixtures containing significant amounts of both heavy and light gas molecules (like WF_6 and H_2), the accurate evaluation of thermal diffusion coefficients may be important and approximations may lead to large errors. Figure 2.5 shows the value of the $\mathbb{D}_i^T$ in a 80 mole% H_2 + 10 mole% WF_6 + 10 mole% SiH_4 gas mixture.

2.8 Numerical solution methods for CVD model equations

The mathematical model described in the previous sections consists of a number of coupled non-linear partial differential equations with boundary conditions, which cannot be solved analytically in general. Therefore, numerical methods have to be applied in order to find approximate solutions of the equations.

For the type of computational fluid dynamics problems as appearing in CVD modeling, three classes of numerical methods are generally used, each of which, in fact, has been applied rather successfully in CVD modeling: (1) Finite element methods, which are especially useful for modeling complex geometries, have mainly been used for the modeling of highly diluted CVD gas systems [*e.g.* 2.12, 2.22-2.27]. (2) Finite difference methods have mainly been used for the numerical solution of complex CVD chemistry models in combination with relatively simple hydrodynamic models [*e.g.* 2.9-2.11, 2.28-2.33]. (3) The finite volume method has been used for modeling 2D and 3D transport phenomena in diluted and undiluted CVD systems [*e.g.* 2.34-2.42]. Published models for tungsten CVD have been based on the finite difference method [2.43] and the finite volume method [2.44-2.48]. A comparison of the advantages and disadvantages of the three methods can be found in [2.49, 2.50].

2.9 The finite volume method for solving CVD model equations

For the tungsten CVD modeling studies described in chapters 4 and 5 of this book, the finite volume method has been used. This method is described in detail in [2.51] and in numerous other publications. Here we will give a brief description of the method only and we will describe in detail how the method has to be adapted in order to include multicomponent diffusion phenomena, thermal diffusion and surface reactions.

2.9.1   Discretization of the general transport equation

The equations (2.1-2.4, 2.30/2.31) in the mathematical model are of the convection-diffusion type and can be written in the general form:

$$\underset{transient}{\xi_\phi \tfrac{\partial}{\partial t}(\rho\phi)} \;=\; \underset{convection}{-\xi_\phi \nabla\cdot(\rho\underline{v}\phi)} \;+\; \underset{diffusion}{\nabla\cdot(\Gamma_\phi\nabla\phi)} \;+\; \underset{source}{S_\phi} \tag{2.97}$$

Here, ϕ is the variable to be solved. The meaning of the factor ξ_ϕ, the diffusion coefficient Γ_ϕ and the source term S_ϕ for each of the equations is explained in table 2.6. Since all the transport equations can be written in the above general form, we can use one numerical algorithm to solve all equations. For simplicity, the numerical method will be explained for the 2D cartesian case. The extension to 2D axisymmetric and 3D situations is straightforward. In 2D cartesian form we find:

$$\xi_\phi\frac{\partial}{\partial t}(\rho\phi) \;=\; -\xi_\phi\frac{\partial}{\partial x}(\rho v_x\phi) - \xi_\phi\frac{\partial}{\partial y}(\rho v_y\phi)$$
$$+\frac{\partial}{\partial x}\left(\Gamma_\phi\frac{\partial\phi}{\partial x}\right) + \frac{\partial}{\partial y}\left(\Gamma_\phi\frac{\partial\phi}{\partial y}\right) + S_\phi$$

The solution domain is divided into a number of adjoining rectangular control volumes, or grid cells, each of them surrounding one grid point in which all scalar variables (pressure P, temperature T, species mass fractions ω_i and fluid properties) are calculated. The vector quantities (velocity $\underline{v}$, and species diffusion fluxes $\underline{j}_i^C,\underline{j}_i^T$) are calculated in points located at the cell walls, halfway between the scalar grid points, using a so called staggered grid. This is illustrated in figure 2.6. Consider the control volume surrounding the grid point P. In the 2D case, P has four neighboring grid points, indicated by N(orth), S(outh), E(ast) and W(est). The corresponding walls of the control volume are indicated by n, s, e and w. For the sake of clarity, we write $v_x = u$ and $v_y = v$ in the remainder of this chapter. Now, by integration of the equation 2.97 over the control volume $\Delta x\Delta y$ surrounding P we find

$$\int\limits_{\Delta x}\int\limits_{\Delta y}(\xi_\phi\frac{\partial\rho\phi}{\partial t})dxdy = \int\limits_{\Delta x}\int\limits_{\Delta y}\left(-\xi_\phi\frac{\partial}{\partial x}(\rho u\phi) - \xi_\phi\frac{\partial}{\partial y}(\rho v\phi)\right.$$
$$\left. +\frac{\partial}{\partial x}\left(\Gamma_\phi\frac{\partial\phi}{\partial x}\right) + \frac{\partial}{\partial y}\left(\Gamma_\phi\frac{\partial\phi}{\partial y}\right) + S_\phi\right)dxdy \tag{2.98}$$

By using the Gauss theorem, assuming that ξ_ϕ prevails over the control

Table 2.6: Terms in the general transport equations

equation	ϕ	ξ_ϕ	Γ_ϕ	S_ϕ
Continuity (2.1)	1	1	0	0
Momentum (2.2) + (2.3)	$\underline{v}$	1	μ	$\nabla \cdot \left(\mu(\nabla \underline{v})^\dagger + (\kappa - \tfrac{2}{3}\mu)(\nabla \cdot \underline{v})\underline{\underline{I}} \right) - \nabla P$ $+\rho \underline{g}$
Energy (2.4)	T	c_p	λ	$\nabla \cdot \left(RT \sum\limits_{i=1}^{N} \frac{\mathbf{D}_i^T}{m_i} \nabla(\ln f_i) \right)$ $+ \sum\limits_{i=1}^{N} \left(\frac{H_i}{m_i} \nabla \cdot \underline{j}_i - \sum\limits_{k=1}^{K} H_i \nu_{ik}(\mathcal{R}_k^g - \mathcal{R}_{-k}^g) \right)$
Species (appr., 2.30)	ω_i	1	$\rho \mathbb{D}_i'$	$\nabla \cdot (\mathbb{D}_i^T \nabla(\ln T)$ $+ m_i \sum\limits_{k=1}^{K} \nu_{ik} \left(k_k \prod\limits_{i=1}^{N} c_i^{\|-\nu_{ik}\|} - k_{-k} \prod\limits_{i=1}^{N} c_i^{\|\nu_{ik}\|} \right)$
Species (exact, 2.31)	ω_i	1	$\rho \mathbb{D}_i$	$\nabla \cdot (\mathbb{D}_i^T \nabla(\ln T) + \nabla \cdot (\rho \omega_i \mathbb{D}_i \nabla(\ln m)$ $-\nabla \cdot \left(m\omega_i \mathbb{D}_i \sum\limits_{j=1}^{N} \frac{j_j}{m_j D_{ij}} \right)$ $+ m_i \sum\limits_{k=1}^{K} \nu_{ik} \left(k_k \prod\limits_{i=1}^{N} c_i^{\|-\nu_{ik}\|} - k_{-k} \prod\limits_{i=1}^{N} c_i^{\|\nu_{ik}\|} \right)$

volume and by linearising the source term S_ϕ in ϕ according to $S_\phi = S_\phi^C + S_\phi^P \phi$ we may write

$$\xi_\phi \int\limits_{\Delta x} \int\limits_{\Delta y} \frac{\partial \rho\phi}{\partial t}\, dxdy = -\xi_\phi \left(\int\limits_n \rho v \phi dx - \int\limits_s \rho v \phi dx + \int\limits_e \rho u \phi dy - \int\limits_w \rho u \phi dy \right)$$

$$+ \int\limits_n \Gamma_\phi \frac{\partial \phi}{\partial y} dx - \int\limits_s \Gamma_\phi \frac{\partial \phi}{\partial y} dx + \int\limits_e \Gamma_\phi \frac{\partial \phi}{\partial x} dy - \int\limits_w \Gamma_\phi \frac{\partial \phi}{\partial x} dy$$

$$+ \int\limits_{\Delta x} \int\limits_{\Delta y} (S_\phi^C + S_\phi^P \phi_P) dxdy \qquad (2.99)$$

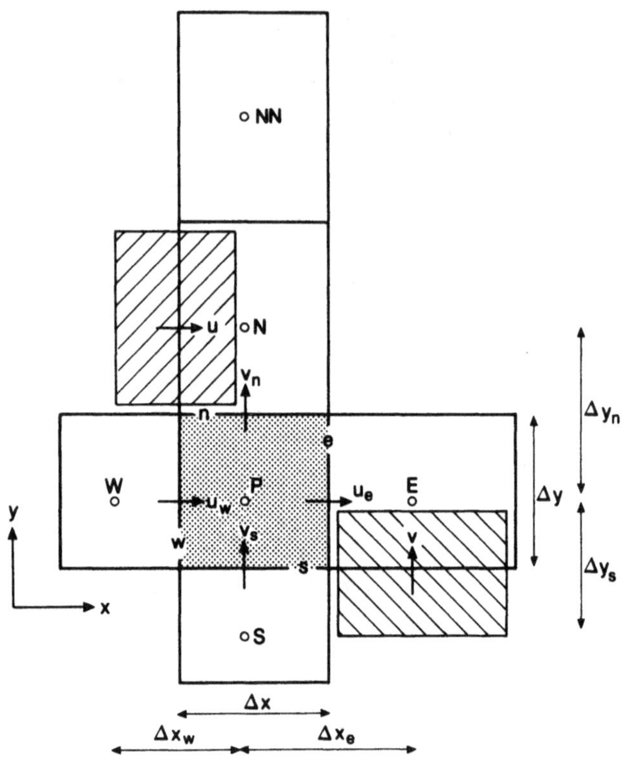

Figure 2.6: Grid cells and staggered grid

Now, the remaining integrals are approximated as

$$
\xi_{\phi,P} \; \frac{\rho_P \phi_P - \rho_P^0 \phi_P^0}{\Delta t} \Delta x \Delta y =
$$
$$
-\quad \xi_{\phi,P} \left(\rho_n v_n \phi_n \Delta x - \rho_s v_s \phi_s \Delta x + \rho_e u_e \phi_e \Delta y - \rho_w u_w \phi_w \Delta y \right)
$$
$$
+\quad \Gamma_{\phi,n} \left. \frac{\partial \phi}{\partial y}\right|_n \Delta x - \Gamma_{\phi,s} \left.\frac{\partial \phi}{\partial y}\right|_s \Delta x + \Gamma_{\phi,e}\left.\frac{\partial \phi}{\partial x}\right|_e \Delta y - \Gamma_{\phi,w}\left.\frac{\partial \phi}{\partial x}\right|_w \Delta y
$$
$$
+\quad (S_\phi^C + S_\phi^P \phi_P)\Delta x \Delta y \tag{2.100}
$$

where the superscript ^{0} indicates the value at the previous time step. Here, a first order backward time differencing has been used, which has first order accuracy in Δt and which is unconditionally stable for all Δt. The densities and diffusion coefficients at the cell walls are approximated as

$$\rho_n = \frac{1}{2}(\rho_P + \rho_N) \tag{2.101}$$

$$\Gamma_{\phi,n} = \frac{2\Gamma_P\Gamma_N}{\Gamma_P + \Gamma_N} \tag{2.102}$$

and similar for other walls. For the approximation of the value of the variable ϕ and of its first derivative on the cell walls several discretization schemes can be used, *e.g.*:

$$\textit{central scheme} \quad : \quad \phi_n \quad = \frac{1}{2}(\phi_P + \phi_N)$$

$$\left.\frac{\partial\phi}{\partial y}\right|_n = \frac{\phi_N - \phi_P}{\Delta y_n} \tag{2.103}$$

$$1^{st}\textit{ order upwind scheme} \quad : \quad \phi_n \quad = \begin{cases} \phi_P \textit{ for } v_n \geq 0 \\ \phi_N \textit{ for } v_n < 0 \end{cases}$$

$$\left.\frac{\partial\phi}{\partial y}\right|_n = \frac{\phi_N - \phi_P}{\Delta y_n} \tag{2.104}$$

$$\textit{hybrid scheme} \quad : \quad \phi_n \quad = \begin{cases} \phi_P & \textit{for } Pe_{c,n} > 2 \\ \frac{1}{2}(\phi_P + \phi_N) & \textit{for } |Pe_{c,n}| \leq 2 \\ \phi_N & \textit{for } Pe_{c,n} < -2 \end{cases}$$

$$\left.\frac{\partial\phi}{\partial y}\right|_n = \begin{cases} \frac{\phi_N - \phi_P}{\Delta y_n} & \textit{for } |Pe_{c,n}| \leq 2 \\ 0 & \textit{for } |Pe_{c,n}| > 2 \end{cases} \tag{2.105}$$

with the cell Peclet number on the n-wall defined as $Pe_{c,n} = \rho_n v_n \Delta y_n/\Gamma_{\phi,n}$, and similar for other walls. The central scheme has second order accuracy in Δx and Δy, but becomes unstable for large $| Pe_c |$, showing wiggles in the solution. The upwind scheme damps these wiggles, in turn for loss of accuracy: this scheme gives only first order accuracy and has a large numerical diffusion. The hybrid scheme is identical with the central scheme for $| Pe_c |\leq 2$. For $| Pe_c |> 2$, the hybrid scheme locally switches to the first order upwind scheme and sets the diffusion contribution to zero. Thus, the hybrid scheme combines some of the advantages of the first order upwind scheme (stability) and the central scheme (accuracy). Because of the low Reynolds numbers, and thus low cell-Peclet numbers, the hybrid scheme usually results in a central differencing scheme for CVD flow simulations.

Using the hybrid differencing scheme and defining convection terms and diffusion terms according to $C_n = \xi_{\phi,P}\rho_n v_n \Delta x$ and $D_n = \Gamma_{\phi,n}\Delta x/\Delta y_n$ (and similar for other walls), we can write *eq.* 2.100 as:

$$a_P\phi_P = a_N\phi_N + a_S\phi_S + a_E\phi_E + a_W\phi_W + b \tag{2.106}$$

where

$$
\begin{aligned}
a_N &= 0 & &for\ Pe_{c,n} > 2 \\
a_N &= D_n - \tfrac{1}{2}C_n & &for\ |Pe_{c,n}| \leq 2 \\
a_N &= -C_n & &for\ Pe_{c,n} < -2
\end{aligned}
\tag{2.107}
$$

$$
\begin{aligned}
a_S &= C_s & &for\ Pe_{c,s} > 2 \\
a_S &= D_s + \tfrac{1}{2}C_s & &for\ |Pe_{c,s}| \leq 2 \\
a_S &= 0 & &for\ Pe_{c,s} < -2
\end{aligned}
\tag{2.108}
$$

and similar for the other walls, and

$$
a_P = a_N + a_S + a_E + a_W - S_\phi^P \Delta x \Delta y + \frac{\xi_{\phi,P}\rho_P^0 \Delta x\ \Delta y}{\Delta t}
\tag{2.109}
$$

$$
b = S^C \Delta x \Delta y + \frac{\xi_{\phi,P}\rho_P^0 \phi_P^0 \Delta x\ \Delta y}{\Delta t}
\tag{2.110}
$$

2.9.2   The SIMPLE algorithm

The above discretization procedure was deduced for a scalar variable on a scalar grid point. A similar procedure can be followed for the discretization of each of the components of the Navier-Stokes equations on the staggered grid points. There is however one fundamental difficulty in the calculation of the velocity field, related to the appearance of the (unknown) pressure gradient in the Navier-Stokes equation. There is no obvious equation for obtaining the pressure distribution. However, the pressure field is indirectly specified through the continuity equation: A correct pressure field will result in a velocity field which satisfies the continuity equation.

A well known method to solve this problem is the so called SIMPLE algorithm (*S*emi *I*mplicit *M*ethod for *P*ressure *L*inked *E*quations). This algorithm has been described in numerous publications [*e.g.* 2.50, 2.51] and we will not discuss it in detail here. Basically, SIMPLE is an iterative procedure, in which the Navier-Stokes equations are first solved for a guessed pressure distribution. The resulting velocity field will in general not satisfy the continuity equation. The deviation from continuity is used to find a correction for the pressure field. For this purpose, a pressure correction equation is solved, which has the same basic form as the general convection-diffusion equation. From the pressure correction a velocity correction is obtained. The corrected pressures and velocities are used as an initial guess for the next iteration. If the procedure converges, the obtained velocity and pressure will approach the correct velocity and pressure, satisfying both the momentum and the continuity equation.

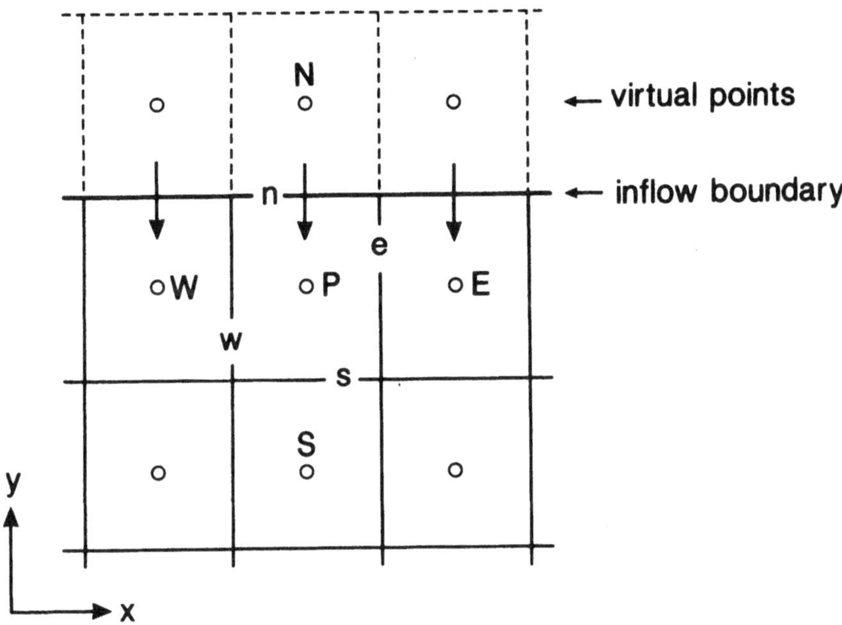

Figure 2.7: Inflow boundary condition

2.9.3 Boundary conditions

For every transport equation of the form of *eq.* 2.97 a set of discretized boundary conditions is needed. These boundary conditions may be imposed through the use of so called virtual grid points outside the computational domain, or by modification of coefficients and source terms in near-boundary grid cells. These methods have been described elsewhere [*e.g.* 2.50, 2.51] and we will not go into details here. Instead, the methods are illustrated for two types of boundary conditions which are typical for CVD reactor modeling: the boundary conditions for the species concentrations in the inflow and the boundary conditions for the normal flow and the species concentrations on reacting walls.

Inflow boundary conditions
In figure 2.7 an inflow opening normal to the y-direction has been illustrated. The computational domain is extended across the inflow boundary with a virtual point. For the species concentration equation we wish to im-

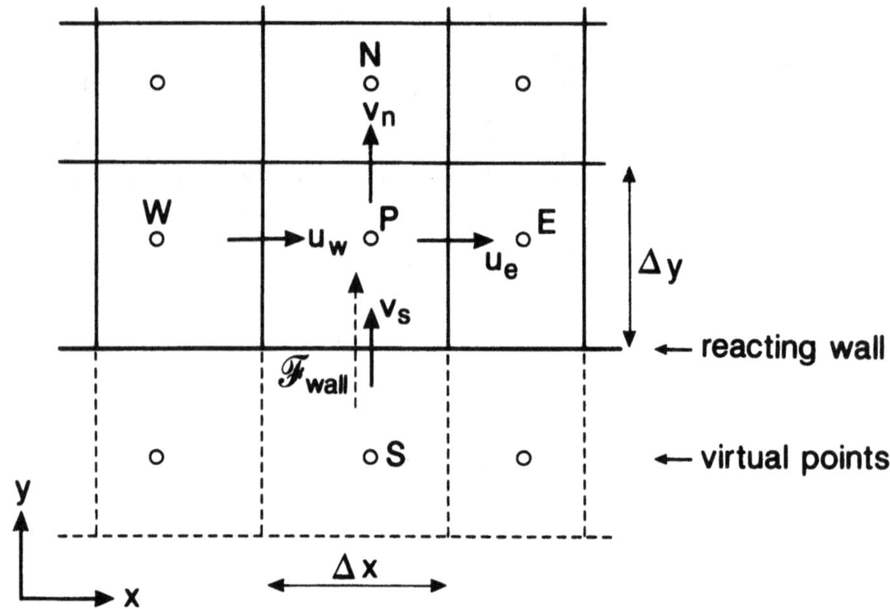

Figure 2.8: Reacting wall boundary condition

pose the boundary conditions (2.51-2.52) on the n-wall of the control volume surrounding P, *i.e.* we want to set the diffusion fluxes (including thermal diffusion) through the inflow opening to zero and we want to give each of the species concentrations in the inflow opening a fixed value $\omega_i = \omega_{i,in}$. The first is simply done by setting the diffusion coefficients in the virtual N-point equal to zero, $\mathbb{D}_{i,N} = \mathbb{D}^T_{i,N} = 0$. Through *eq.* 2.102 we now find zero diffusion coefficients on the n-wall of the cell, $\mathbb{D}_{i,n} = \mathbb{D}^T_{i,n} = 0$, and through *eqs.* 2.15 and 2.19 we then find $j^C_{iy,n} = j^T_{iy,n} = 0$. The second can be achieved by setting $\omega_{i,N} = \omega_{i,in}$. Since $\mathbb{D}_{i,n} = 0$, the hybrid differencing scheme, through *eq.* 2.105, will then lead to $\omega_{i,n} = \omega_{i,N} = \omega_{i,in}$. Note that the zero diffusion inflow boundary condition, which is not commonly used in the modeling of fluid flow with heat and mass transfer, is especially important in CVD modeling because of the low Reynolds (and thus low Peclet) numbers common in CVD, causing a diffusive species transport through the inflow opening which is significant relative to the convective transport.

Boundary conditions at a reacting wall

In figure 2.8 we have illustrated a boundary formed by a reacting wall (*e.g.* a wafer surface). For the normal velocity on the s-wall of the control volume surrounding P we want to impose the boundary condition 2.42, *i.e.*

$$v_s = \frac{1}{\rho} \sum_{i=1}^{N} m_i \sum_{l=1}^{L} \sigma_{il} \mathcal{R}_l^S \tag{2.111}$$

This is simply done by setting v_s to the desired value (no transport equation for v_s is solved). For the species concentration equations we want to impose *eq.* 2.45, *i.e.*

$$\rho \omega_i v_s + j_{iy,s}^C + j_{iy,s}^T = -m_i \sum_{l=1}^{L} \sigma_{il} \mathcal{R}_l^S \tag{2.112}$$

The right-hand term in *eq.* 2.112 is found by evaluating the surface reaction rates $\mathcal{R}^S$ at the temperature, the pressure and the species concentrations in the grid point P from the previous iteration. For brevity, we write

$$- m_i \sum_{l=1}^{L} \sigma_{il} \mathcal{R}_l^S (T_P, P_P, \omega_{j,P}(j = 1, N)) = \mathcal{F}_{wall} \tag{2.113}$$

where $\mathcal{F}_{wall}$ is the mass flux of the ith species deposited on the s-wall of the control volume. When we now equate the sum of the convective and diffusive fluxes through the s-wall to $\mathcal{F}_{wall}$ and we write a balance equation over the control volume surrounding P similar to *eq.* 2.100, we get

$$\begin{aligned}
\frac{\rho_P \omega_P - \rho_P^0 \omega_P^0}{\Delta t} &\Delta x \Delta y \\
&= \rho_n v_n \omega_n \Delta x + \rho_e u_e \omega_e \Delta y - \rho_w u_w \omega_w \Delta y \\
&+ \left. \Gamma_{\phi,n} \frac{\partial \omega}{\partial y} \right|_n \Delta x + \left. \Gamma_{\phi,e} \frac{\partial \omega}{\partial x} \right|_e \Delta y - \left. \Gamma_{\phi,w} \frac{\partial \omega}{\partial x} \right|_w \Delta y \\
&+ \mathcal{F}_{wall} \Delta x + S \Delta x \Delta y
\end{aligned} \tag{2.114}$$

2.9.4   Stefan-Maxwell equations and thermal diffusion

In the general convection-diffusion equation 2.97 it is assumed, that diffusive transport of the quantity ϕ is due to a gradient in ϕ only. For species diffusion in a multi-component gas mixture this is not true: the diffusive flux of a species in a N component mixture depends on the concentration gradients of all species. One strategy that could be followed for treating multicomponent ordinary diffusion is to set Γ_ϕ in *eq.* 2.97 equal to zero and to include the $\nabla \cdot j_i^C$ term in the "source" term S_ϕ. The diffusive mass flux vectors j_i^C ($i=1,N$) may then be found through the solution of the matrix equation

$$\frac{m}{\rho}\begin{pmatrix} -\sum_{j\neq 1}\frac{\omega_j}{m_j D_{1j}} & \frac{\omega_1}{m_2 D_{12}} & \cdot & \frac{\omega_1}{m_N D_{1N}} \\ \frac{\omega_2}{m_1 D_{21}} & -\sum_{j\neq 2}\frac{\omega_j}{m_j D_{2j}} & \cdot & \frac{\omega_2}{m_N D_{2N}} \\ \cdot & \cdot & & \cdot \\ \cdot & \cdot & \cdot & \\ \frac{\omega_N}{m_1 D_{N1}} & \frac{\omega_N}{m_2 D_{N2}} & \cdot & -\sum_{j\neq N}\frac{\omega_j}{m_j D_{Nj}} \end{pmatrix}\begin{pmatrix} j_1^C \\ j_2^C \\ \cdot \\ \cdot \\ j_N^C \end{pmatrix} =$$

$$\begin{pmatrix} \nabla\omega_1 + \omega_1\nabla(\ln m) \\ \nabla\omega_2 + \omega_2\nabla(\ln m) \\ \cdot \\ \cdot \\ \nabla\omega_N + \omega_N\nabla(\ln m) \end{pmatrix} \qquad (2.115)$$

which is equivalent with *eq.* 2.12. This strategy, however, would dramatically change the character of the general convection-diffusion equation and of the numerical scheme, deteriorating the rate of convergence of the iterative solution procedure. Moreover, the inversion of the $N \times N$ matrix equation 2.115 in each grid point demands a great computational effort.

Alternatively, a strategy can be used in which the diffusive mass flux j_i^C is written in a form which closely resembles an ordinary gradient diffusion term and which has led to *eq.* 2.15:

$$j_i^C = -\rho \mathbb{D}_i \nabla\omega_i - \rho\omega_i \mathbb{D}_i \nabla(\ln m) + m\omega_i \mathbb{D}_i \sum_{\substack{j=1 \\ j\neq i}}^{N} \frac{j_i^C}{m_j D_{ij}} \qquad (2.116)$$

or, in the 2D cartesian case

$$j_{ix}^C = -\rho \mathbb{D}_i \frac{\partial\omega_i}{\partial x} - \rho\omega_i \mathbb{D}_i \frac{\partial\ln m}{\partial x} + m\omega_i \mathbb{D}_i \sum_{\substack{j=1 \\ j\neq i}}^{N} \frac{j_{ix}^C}{m_j D_{ij}} \qquad (2.117)$$

$$j_{iy}^C = -\rho \mathbb{D}_i \frac{\partial\omega_i}{\partial y} - \rho\omega_i \mathbb{D}_i \frac{\partial\ln m}{\partial y} + m\omega_i \mathbb{D}_i \sum_{\substack{j=1 \\ j\neq i}}^{N} \frac{j_{iy}^C}{m_j D_{ij}} \qquad (2.118)$$

Here, the first terms on the right hand sides are of the gradient diffusion type and can be treated in the usual way as the diffusion term in the general convection-diffusion equation 2.97. The last two terms on the right hand sides, which generally are smaller than the first, are included in the "source" term S_ϕ. Their values are evaluated at the previous iteration level. For this purpose, the values of each of the components of the $\underline{j}_i^C$ vectors have to be known. Therefore, after updating the species concentration values, the values of $\underline{j}_{ix}^C$ and $\underline{j}_{iy}^C$ are calculated from *eqs.* 2.117 and 2.118 and stored. As was done with the velocity components, the components of the $\underline{j}_i^C$ vectors are calculated on the walls of the scalar grid cells.

The terms associated with the thermal diffusion fluxes in *eqs.* 2.30 and 2.31 can be included in the source term S_ϕ of the general transport equation. In general, these terms vary slowly during the iteration process. Therefore, their value does not have to be updated every iteration, allowing a substantial saving of computer effort, especially when the full multicomponent formulation for calculating $\mathbb{D}^T$ is used.

2.9.5 Iterative solution procedures

The discretized transport equations give rise to a matrix equation of the form $\underline{\underline{A}} \, \underline{\phi} = \underline{b}$, where the matrix $\underline{\underline{A}}$ contains the coefficients a from *eq.* 2.106, $\underline{\phi}$ is a vector representing all the unknown variables in all grid points and $\underline{b}$ is a vector representing the source terms b in all grid points. Through the fluid properties and the velocities, both $\underline{\underline{A}}$ and $\underline{b}$ are functions of $\underline{\phi}$ and the matrix equation is nonlinear. Therefore, it has to be linearized and solved iteratively. After linearization, a direct solution procedure could in principle be applied, but the size of the matrix $\underline{\underline{A}}$ makes direct solution procedures prohibitive in computational effort. Also, direct solution procedures have a small range of convergence, necessitating an accurate initial guess. Therefore, iterative solution procedures may be advantageous, not only to deal with the nonlinearities, but also for the solution of the linearized matrix equation. The iterative nature of the solution procedure is then introduced on three levels: *(i)* First, the transport equations for the different variables are decoupled. Thus, a set of matrix equations $\underline{\underline{A}}_i \underline{\phi}_i = \underline{b}_i$ is obtained, where $\underline{\phi}_i$ is a vector representing the values of the ith variable in all grid points and $\underline{\underline{A}}_i$ and $\underline{b}_i$ contain the coefficients and source terms for this variable. The coupling between the different $\underline{\phi}_i$ is now accounted for through the repeated, iterative solution of the decoupled equations for all $\underline{\phi}_i$. *(ii)* The decoupled equations still contain non-linearities, which are linearized by evaluating the values of the coefficients in $\underline{\underline{A}}_i$ and the source terms in $\underline{b}_i$ at the previous iteration level. *(iii)* The resulting, linearized matrix equations are also solved iteratively.

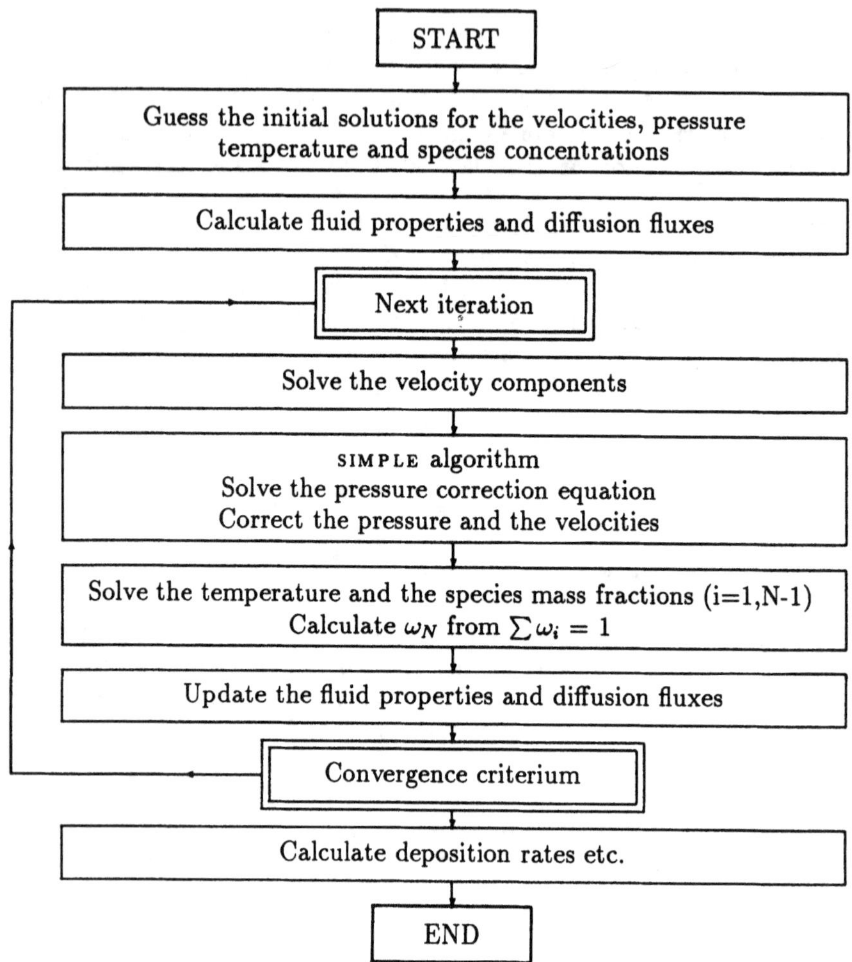

Figure 2.9: Iterative solution procedure for stationary calculations

For this purpose, iterative solvers such as point-by-point or line-by-line Gauss-Seidel methods can be used. Furthermore, iteration is used to couple the Navier-Stokes and continuity equations in the SIMPLE procedure, and for the solution of the Stefan-Maxwell equations through *eq.* 2.116. For clarity, we will now give an overview of the total procedure:

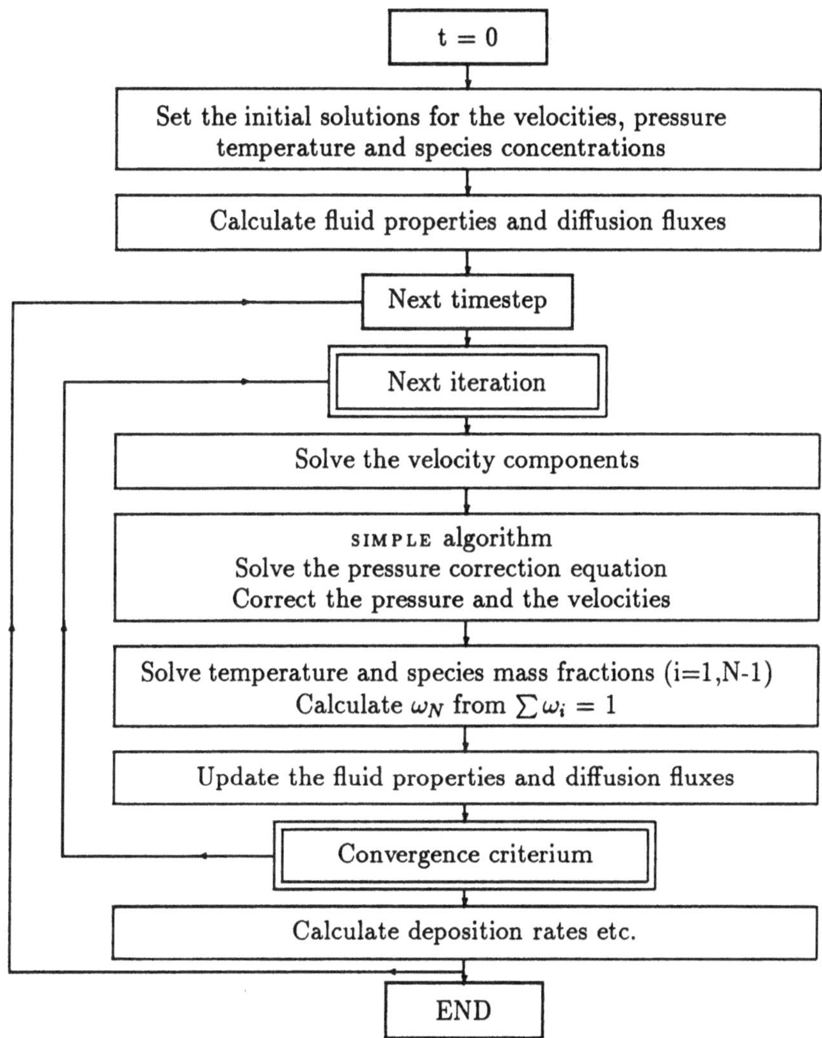

Figure 2.10: Iterative solution procedure for instationary calculations

For stationary calculations, the iterative procedure is illustrated in figure 2.9. We start by making an initial guess for the values of all primary variables (velocity components, pressure, temperature and species concentrations) in all grid points, and we calculate the corresponding fluid properties. Then the iterative solution procedure is started. We first solve the velocity components in all grid points. Then the pressure correction equation is solved and the velocities and pressures are corrected through the SIMPLE algorithm. Finally, the energy equation and the species concentration equations for all species except one are solved. At the end of the iteration, the fluid properties and diffusion fluxes are updated and a number of convergence criteria are checked. If these criteria are not fulfilled, the above procedure is repeated.

For transient calculations, the procedure is illustrated in figure 2.10. We now have to know the initial fields at $t = 0$. From here, t is increased with a time step Δt and the solution at the new time level is found iteratively. When the solution at a time level fulfills the convergence criteria, we move to the next time step.

In order for the iterative solution procedure to converge, an underrelaxation of the equations is needed. For this purpose the values ϕ_k^* obtained at a certain iteration level k are used only partially to update the values ϕ_{k-1} from the previous iteration level: $\phi_k = \eta \phi_k^* + (1 - \eta)\phi_{k-1}$, where the relaxation factor $\eta < 1$ causes underrelaxation. Typical values for η are $\eta = 0.1\text{-}0.8$. When gas phase reactions play an important role in the species concentration equations the above relaxation method may not be adequate, due to the so called "stiffness" of the equations. Better results may be obtained by limiting the maximum relative change in the concentration between one iteration and the next to a small value Δ, with Δ typically of the order of 0.01-0.1 [2.42, 2.52].

When the iterative solution procedure is converging to the desired solution, the iterations can be stopped when the difference between the iterated solution and the final solution is small enough. Several criteria for stopping the iterative procedure should be checked, *e.g.* *(i)* The value of the residuals R_ϕ, averaged over all grid points, for each variable ϕ. Here, R_ϕ is defined as $R_\phi = |\, a_P\phi_P - a_E\phi_E - a_W\phi_W - a_N\phi_N - a_S\phi_S - b\,|$. *(ii)* The relative change from one iteration to another for all variables in all grid cells, and *(iii)* The error in the overall energy balance and each of the overall species mass balances.

References Chapter 2

[2.1]   G.A. Bird (1976) "Molecular Gas Dynamics", Clarendon Press, Oxford, United Kingdom

[2.2]   R.B. Bird, W.E. Stewart and E.N. Lightfood (1960), "Transport Phenomena", John Wiley & Sons, New York, USA

[2.3]   R. Clark Jones and W.H. Furry (1946), "The Separation of Isotopes by Thermal Diffusion", *Rev. of Modern Phys.* **18** (2), pp. 151-224

[2.4]   K.E. Grew and T.L. Ibbs (1952), "Thermal Diffusion in Gases", Cambridge Univ. Press, Cambridge, Great Britain

[2.5]   J.O. Hirschfelder, C.F. Curtiss and R.B. Bird (1967), "Molecular Theory of Gases and Liquids", John Wiley and Sons Inc., New York, USA

[2.6]   D.D. Wagman, W.H. Evans, V.B. Parker, R.H. Schumm, I. Halow, S.M. Bailey, K.L. Churney and R.L. Nuttall (1982), "The NBS Tables of chemical thermodynamic properties", *J. Phys. Chem. Ref. Data* **11**, Suppl. 2

[2.7]   "JANAF Thermochemical tables" (1985), Dow Chemical Company

[2.8]   I. Barin and O. Knacke (1973), "Thermochemical properties of inorganic substances", Springer, Berlin, Germany

[2.9]   M.E. Coltrin, R.J. Kee and J.A. Miller (1984), "A mathematical model of the coupled fluid mechanics and chemical kinetics in a chemical vapor deposition reactor", *J. Electrochem. Soc.* **131** (2), pp. 425-434

[2.10] M.E. Coltrin, R.J. Kee and J.A. Miller (1986), "A mathematical model of silicon chemical vapor deposition. Further refinements and the effects of thermal diffusion", *J. Electrochem. Soc.* **133** (6), pp. 1206-1213

[2.11] M.E. Coltrin, R.J. Kee and G.H. Evans (1989), "A mathematical model of the fluid mechanics and gas-phase chemistry in a rotating disk chemical vapor deposition reactor", *J. Electrochem. Soc.* **136** (3), pp. 819-829

[2.12] H.K. Moffat and K.F. Jensen (1988), "Three-dimensional flow effects in silicon CVD in horizontal reactors", *J. Electrochem. Soc.* **135** (2), pp. 459-471

[2.13] A. Sherman (1988), "Modeling of Chemical Vapor Deposition Reactors", *J. Electr. Mat.* **17** (5), pp. 413-423

[2.14] l'Air Liquide, Division Scientifique (1976), "Encyclopédie des

Gaz", Elseviers Scientific Publ. Comp., Amsterdam, The
Netherlands

[2.15] G.C. Maitland and E.B. Smith (1972), "Critical Reassessment of
Viscosities of 11 Common gases", *J. Chem. and Eng. Data* **17**
(2), pp. 150-156

[2.16] R.C. Weast (*ed.*) (1984), "Handbook of Chemistry and Physics",
CRC Press Inc., Boca Raton, Florida, USA

[2.17] S. Bretsznajder (1971), "Prediction of Transport and other
Physical Properties of Fluids", (translated by J. Bandrowski),
Pergamon Press, Oxford, Great Britain

[2.18] R.C. Reid, J.M. Prausnitz and B.E. Poling, (1987), "The
Properties of Gases and Liquids", (2nd edition), McGraw-Hill,
New York

[2.19] R.A. Svehla (1962), "Estimated Viscosities and Thermal
Conductivities of Gases at high Temperatures", NASA
Technical Report R-132

[2.20] C.R. Wilke and C.Y. Lee (1955), *Ind. Eng. Chem.*, **47**, pp. 1253

[2.21] R. Clark Jones (1941), "On the theory of the Thermal Diffusion
Coefficient for Isotopes II", *Phys. Rev.* **59**, pp. 1019-1033

[2.22] C. Houtman, D.B. Graves and K.F. Jensen (1986), "CVD in
stagnation point flow. An Evaluation of the classical 1D
treatment", *J. Electrochem. Soc.* **133** (5), pp. 961-970

[2.23] P. Lee, D. McKenna, D. Kapur and K.F. Jensen (1986),
"MOCVD in inverted stagnation point flow. I. Deposition of
GaAs from TMAs and TMGa", *J. Crystal Growth* **77**,
pp. 120-127

[2.24] H.K. Moffat and K.F. Jensen (1986), "Complex flow phenomena
in MOCVD reactors, I: Horizontal reactors", *J. Crystal Growth*
77, pp. 108-119

[2.25] D.I. Fotiadis, A.M. Kremer, D.R. McKenna and K.F. Jensen
(1987) "Complex flow phenomena in vertical MOCVD reactors:
effects on deposition uniformity and interface abruptness",
J. Cryst. Growth **85**, pp. 154-164

[2.26] D.I. Fotiadis, S. Kieda and K.F. Jensen (1990), "Transport
phenomena in vertical reactors for metalorganic vapor phase
epitaxy: I. effects of heat transfer characteristics, reactor
geometry, and operating conditions", *J. Cryst. Growth* **102**,
pp. 441-470

[2.27] M. Pons, R. Klein, C. Arena and S. Mariaux (1989), "Modeling
of cold wall chemical vapor deposition reactors (for
semiconductor fabrication)", *J. de Physique*, Coll. C5,

Suppl. au no. 5, Tome 50, pp. c5-57-65

[2.28] R. Pollard and J. Newman (1980), "Silicon deposition on a
rotating disk", *J. Electrochem. Soc.* **127** (3), pp. 744-752

[2.29] M. Michaelidis and R. Pollard (1984), "Analysis of chemical
vapor deposition of boron", *J. Electrochem. Soc.* **131**
(4), pp. 860-868

[2.30] J.P. Jenkinson and R. Pollard (1984), "Thermal diffusion effects
in chemical vapor deposition reactors", *J. Electrochem. Soc.*
131 (12), pp. 2911-2917

[2.31] J. Jůza and J. Cermák (1982), "Phenomenological model of the
CVD epitaxial reactor", *J. Electrochem. Soc.* **129** (7),
pp. 1627-1634

[2.32] H. Chehouani, B. Armas, S. Benet and S. Brunet (1989),
"Simulation du transfert de chaleur et de quantité de mouvement
dans un reacteur de vapodéposition", *J. de Physique*,
Coll. C5, Suppl. au no. 5, Tome 50, pp. c5-47-56

[2.33] C. Vinante, P. Duverneuil J.P. Couderc (1989), "A two
dimensional model for LPCVD reactors hydrodynamics and mass
transfer" *J. de Physique*, Coll. C5, Suppl. au no. 5, Tome
50, pp. c5-57-65

[2.34] G. Evans and R. Greif (1987), "A numerical model of the flow
and heat transfer in a rotating disk chemical vapor deposition
reactor", *J. Heat Transfer* **109**, pp. 928-935

[2.35] G. Evans and R. Greif (1989), "A study of traveling wave
instabilities in a horizontal channel flow with applications to
chemical vapor deposition", *Int. J. Heat Mass Transfer*
(5), pp. 895-911

[2.36] G. Wahl (1977), "Hydrodynamic description of CVD processes",
Thin Solid Films **40**, pp. 13-26

[2.37] J. Ouazzani and F. Rosenberger (1990), "Three-dimensional
modelling of horizontal Chemical Vapor Deposition",
J. Cryst. Growth. **100**, pp. 545-576

[2.38] S.A. Gokoglu, M. Kuczmarski, P. Tsui and A. Chait (1989),
"Convection and chemistry effects in CVD - A 3-D analysis for
silicon deposition", *J. de Physique*, Coll. C5, Suppl. au
no. 5, Tome 50, pp. c5-17-35

[2.39] S. Rhee, J. Szekely and O.J. Ilegbusi (1987), "On three
dimensional transport phenomena in CVD processes",
J. Electrochem. Soc. **134** (10) pp. 2552-2559

[2.40] Ch. Hopfmann, Ch. Werner and J.I. Ulacia F. (1991), "Numerical
analysis of fluid flow and nonuniformities in a polysilicon

LPCVD batch reactor" *Appl. Surf. Science* **52**, pp. 169-187

[2.41] C.R. Kleijn, T.H. van der Meer and C.J. Hoogendoorn (1989), A mathematical model for LPCVD in a single-wafer reactor", *J. Electrochem. Soc.* **136** (11), pp. 3423-3432

[2.42] C.R. Kleijn (1991), "A mathematical model of the hydrodynamics and gas-phase reactions in silicon LPCVD in a single-wafer reactor", *J. Electrochem. Soc.* **138**, pp. 2190-2200

[2.43] R. Arora and R. Pollard (1991) "A mathematical model for Chemical Vapor Deposition influenced by surface reaction kinetics: Application to low pressure deposition of tungsten", *J. Electrochem. Soc.* **138** (5), pp. 1523-1537

[2.44] T.J. Jasinski and S.S. Kang (1991) "Application of numerical modelling for CVD simulation test case: Blanket tungsten deposition uniformity" in "Tungsten and Other Advanced Metals for ULSI Applications in 1990", G.C. Smith and R. Blumenthal (*eds*), pp. 219-230, The Materials Research Society, Pittsburgh, USA

[2.45] E.J. McInerney, P. Geraghty and S. Kang (1990) "Modeling of WF_6 surface concentration and its effects on the step coverage of hydrogen reduced tungsten films" in "Tungsten and Other Advanced Metals for VLSI/ULSI Applications V", S.S. Wong and S. Furukawa (*eds.*), pp. 135-141, The Materials Research Society, Pittsburgh

[2.46] J.I. Ulacia F., S. Howell, H. Körner and Ch. Werner (1989), "Flow and reaction simulation of a tungsten CVD reactor", *Appl. Surf. Science*, **31**, pp. 370-385

[2.47] C.R. Kleijn and C.J. Hoogendoorn; A. Hasper, J. Holleman and J. Middelhoek (1990) "An experimental and modelling study of the tungsten LPCVD growth kinetics from $H_2 - WF_6$ at low WF_6 partial pressures" In "Tungsten and other advanced metals for VLSI/ULSI applications V", S. Wong and S. Furukawa (*eds*), The Materials Research Society, Pittsburgh, USA, pp. 109-116

[2.48] C.R. Kleijn, and C.J. Hoogendoorn; A. Hasper, J. Holleman, J. Middelhoek (1991) "Transport phenomena in tungsten LPCVD in a single-wafer reactor", *J. Electrochem. Soc.* **138**, pp. 509-517

[2.49] T.M. Shih (1984), "Numerical heat transfer. Series in computational methods in mechanics and thermal sciences", Hemisphere Publ. Corp., Washington

[2.50] W.J. Minkowycz, E.M. Sparrow, G.E. Schneider and R.H. Pletcher (1988), "Handbook of numerical heat transfer", John Wiley and Sons, New York, USA

[2.51] S.V. Patankar (1980), "Numerical heat transfer and fluid flow", Hemisphere Publ. Corp., Washington

[2.52] C.R. Kleijn (1991), "Transport Phenomena in Chemical Vapor Deposition Reactors", Ph.D. Thesis, Delft University of Technology, The Netherlands

Chapter 3

Thermal modeling

3.1 Introduction

In the partial differential equations used so far in our treatment, the temperature at each point in the gas was calculated as a solution of the heat transport equation (2.4). The only inputs necessary are the heat conductivity λ, the heat capacity c_p of the gas, and the temperature values on the solid boundaries of the reactor. In section 2.7 we have seen that reliable values for the transport parameters λ and c_p can be obtained using gas kinetic theory and table data for pure gas components.

However the temperature profiles along the inner solid surfaces of the reactor are not at all straightforward to calculate. In the experimental setup some of the walls are cooled or heated to a fixed temperature, so that these values just need to be measured and used as boundary conditions in the calculations. Other walls are not fixed with respect to temperature and they will acquire their temperature by heat exchange with other solid walls, with the gas inside the reactor chamber and also with the ambient outside the reactor. Even a measurement of the temperature profile on those walls would not be sufficient, because the temperature will adjust to changing gas flow or pressure inside the reactor, so that the measurement would have to be repeated for every working condition.

It is the purpose of this chapter to discuss models for solid wall temperature calculation and apply them for two different tungsten CVD reactors.

3.2   Heat transfer mechanisms

There are basically three different ways of heat exchange between solid surfaces: Convection, conduction and radiation. Convective and conductive heat transport is already contained in the heat balance equation (2.4) of the gas inside the reactor and can be similarly applied to the solid parts of the reactor when the transport properties are appropriately adjusted to the values of the solid material (stainless steel, nickel, graphite etc.).

The radiative heat transfer can be modelled by a direct energy exchange between the single surface elements considering their mutual view factors and it depends on the fourth power of the surface element temperature values.

Due to their different dependence on pressure and temperature the three mechanisms of convection conduction and radiation are of variable importance in different working conditions. Nevertheless, it turns out that none of them can be neglected in a typical LPCVD reactor for tungsten deposition.

3.3   Conduction across very small gaps

Though we have made the general assumption in chapter 2 that the gas mixture behaves as a continuum, there is one exception in the heat transfer between the wafer and the heated susceptor. Usually the wafer is clamped to the susceptor with several holding clamps and the distance between the wafer and the susceptor surfaces are below 0.2mm. This is small enough to be neglected in gas flow calculations, so that susceptor and silicon substrate can be regarded as one solid body.

However, at low enough pressure the thermal contact between wafer and susceptor is rather poor, so that a temperature drop of $10°C$ or more can occur. Since the gap between the two surfaces is comparable to or smaller than the molecular mean free path in the gas, modifications to the continuum flow model of heat transfer have to be considered.

For a rigorous solution the continuum equation (2.4) has to be replaced by the Boltzmann transport equation, which could be solved e.g. by a Monte Carlo particle technique [3.3]. However, for rather small deviations from the continuum regime ($0.01 < Kn < 0.2$, where the Knudsen number Kn is defined as the ratio of the mean free path ℓ to the gap distance d), usable first order corrections can be obtained when the equation (2.4) is left unchanged, but the boundary condition (2.38) is changed to [3.2]

$$T = T_{wall} + \beta'\ell\nabla T \cdot \underline{n} \tag{3.1}$$

where $\underline{n}$ is a unity vector normal to the wall surface, ℓ is the molecular mean free path and β' is a constant in the order of unity.

For two parallel plates at a distance of d the heat flux from the hot to the cold plate is then [3.1]

$$J_q = \frac{\lambda(T_h - T_c)}{d + 2\,\beta'\ell} \tag{3.2}$$

where λ is the thermal conductivity of the gas, $T_h - T_c$ is the temperature difference between the hot (T_h) and cold (T_c) surfaces, and d is their distance.

Though this boundary condition (3.1) is only true for $Kn < 0.2$, the integral solution (3.2) for two parallel plate happens to be a good approximation also for $Kn \geq 0.2$. This is, however only true for the parallel plate arrangement and would lead to errors for other geometries, e.g. cylindrical surfaces.

From (3.2), it is possible to identify two distinctive regions for heat conduction as a function of pressure [3.1]. At high pressures, where the mean-free path l of the gas is small compared with the distance between the wafer and the susceptor, the heat conductivity is independent of pressure. This is true for 10 torr and above, and a negligible temperature drop across the gap is calculated. However, at low pressures, the heat conduction is proportional to the mean-free path and, therefore, inversely on pressure. In the limits the heat-conduction flux J_q approaches the following expressions:

$$J_q = \frac{\lambda(T_h - T_c)}{d};\ \ for\ \ \ell \ll d/\beta'$$

$$J_q = \frac{\lambda(T_h - T_c)}{2\,\beta'\ell};\ \ for\ \ \ell \gg d/\beta'$$

Kennard [3.2] has given a correlation between the parameter β' and the thermal accomodation coefficient α [3.1].

$$\beta' = \frac{2 - \alpha}{\alpha} \cdot \frac{9\gamma - 5}{2\gamma + 2} \tag{3.3}$$

with $\gamma = c_p/c_v$ being the ratio of specific heats for constant pressure and constant volume respectively ($\gamma = 1.4$ for H_2). However, we will not attempt to use a priori values for β', but rather take this parameter as an empirical constant, which is determined from measurement fitting.

Moreover also the distance d is not well known since it depends on clamping strength and also on wafer warpage. So equation (3.2) is rather an empirical qualitative model, which needs adjustment to experimental temperature measurements to become quantitive.

3.4   Radiative heat transfer

The radiative heat exchange can lead to significant energy transport between opposing surfaces, when they are at different temperature. In a cold wall CVD reactor at least part of the solid reactor walls are held at a controlled temperature, either by water cooling or by a temperature controlled heating unit. For those walls the temperature enters the simulation as a fixed boundary condition and the radiative heat flux into these walls will only change the power of the heating arrangement to maintain this temperature.

However, at those walls, which are not temperature controlled, we have to take into account the radiative heat exchange to be able to calculate the correct temperature. The net radiative heat flux J_j^{rad} of a solid wall element A_j can be written as

$$J_j^{rad} = \frac{1}{A_j} \sum_k a_j G_{jk} e_k A_k \sigma T_k^4 - e_j \sigma T_j^4 \tag{3.4}$$

Here a_j and e_j are the absorptivity and the emissivity of the surface element A_j, respectively, T_j is its temperature, A_j its area, and $\sigma = 5.6 \cdot 10^{-8}\ W/(m^2 K^4)$ is the Stefan-Boltzmann radiation constant. G_{ij} are the Gebhardt's factors, which are related with the viewfactors F_{ij} by the equation

$$G_{ij} = F_{ij} + \sum_k r_k F_{ik} G_{kj} \tag{3.5}$$

Here r_k is the reflectivity of the element A_j and F_{ij} is the viewfactor between the elements A_i and A_j defined as

$$F_{ij} = 1/A_i \int \frac{cos\Theta_1 cos\Theta_2}{\pi r_{12}^2} dA_1 dA_2 \qquad (3.6)$$

with r_{12} being the distance between two surface points on A_i and A_j and Θ_1 the angle between the surface normal and the connecting line. The sum in Equ. 3.4 has to be performed over all surface elements A_k visible from the element A_j.

In this formulation several approximations are assumed, which are discussed below:

- Complete transparency of the gas, which is very well fulfilled in the low pressure regime, in which LPCVD of tungsten is performed (see section 2.2).

- Grey body approximation, assuming that a_j, e_j, and r_j are not dependent on the wavelength of the emitted radiation. Since these parameters are no unique material constants but depend e.g. on surface polishing, it seems allowed to neglect their spectral dependence.

- Lambert's radiation law, which assumes a cosine distribution of the emitted radiation.

- Diffuse reflection, assuming that the reflected radiation intensity shows a cosine angular distribution regardless of the angle of incoming radiation. This might be true for walls with microscopically rough surfaces.

The main reason for these assumptions is that they are necessary to yield bearable calculation times, and that may be the real reason, why they are so generally accepted in radiation heat transfer problems. However, we did no comparison of our modeling results with more sophisticated models, and hence cannot give a quantitative accuracy for our results. From that fact it is obvious, that only qualitative results can be expected from Equ.3.4.

3.5   Applications

3.5.1   1-D model for the substrate temperature

In the reactor depicted in Fig. 3.1 the temperature control pyrometer is focussed onto the susceptor surface besides the wafer to guarantee heat control independent of the surface film structure on the wafers.

Depending on the pressure the temperature on the wafer may deviate up to 30°C from the susceptor temperature. Moreover, measurements revealed a lateral temperature profile both on the susceptor and on the substrate, which is due to a heat loss through the mechanical susceptor suspension with the chamber walls. Without doing more sophisticated modeling we describe this heat loss as proportional to the temperature

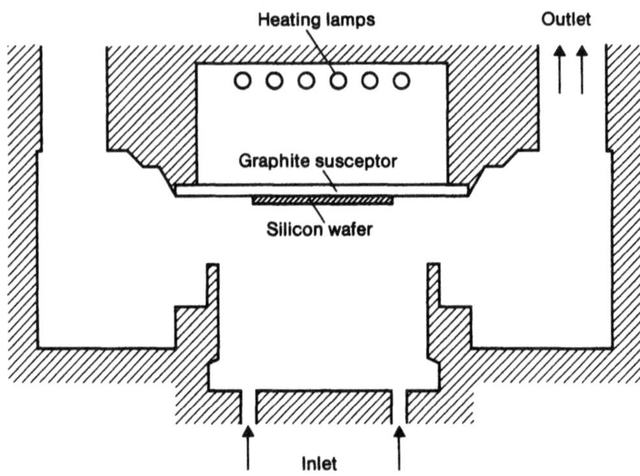

Figure 3.1: Reactor geometry showing the heating lamps, graphite susceptor, silicon wafer and reactor chamber

difference between the holder and the reactor wall T_w as

$$S_L = A(T - T_w) \tag{3.7}$$

where A is a constant that describes the heat transport efficiency.

The different paths of heat exchange are schematically depicted in Fig.3.2. In addition to these elements we have the heat transfer through the gas which is coupled to the solid surface temperatures by means of the boundary conditions.

If we neglect the lateral temperature variations on susceptor, substrate and reactor wall, we can derive a simplified solution for the temperature drop.

In this case the radiation problem consists only of four surface elements, namely the susceptor, front and back side of the silicon substrate, and the cooled reactor walls with temperatures T_s, T_f, T_b, and T_c. The hot susceptor and the cold walls are maintained at fixed values by heating or cooling, while the wafer temperatures have to be determined from the simulation. Due to the high thermal conductivity of silicon, we can assume $T_f = T_b = T_w$.

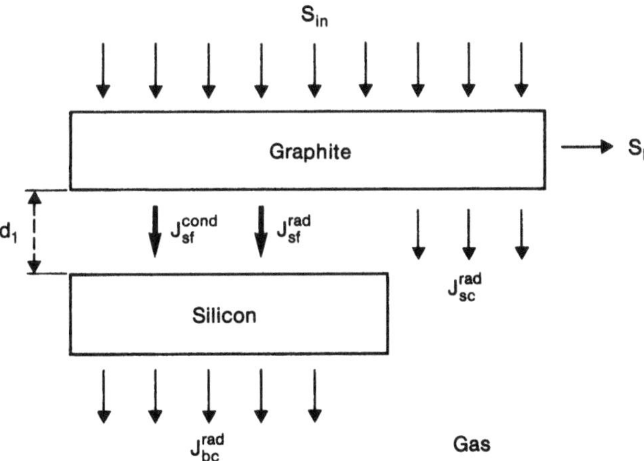

Figure 3.2: Heat sources applied to the wafer, susceptor, and gas. S_{in} is the incoming heat from the lamps (independent of T), S_L is the lateral heat loss through mechanical susceptor suspension and is related to the wall temperature T_w.

In the 1-D model we have the viewfactors

$$F_{sf} = F_{bc} = 1$$

and $F_{ij} = 0$ for all other pairs.

With Equ. (3.5) we can derive the Gebhard's factors G_{ij}, and using Equ. (3.4) we get

$$J_{sf}^{rad} = C_{sf}\sigma(T_s^4 - T_w^4)$$

and

$$J_{bc}^{rad} = C_{bc}\sigma(T_w^4 - T_c^4)$$

with the exchange factor C_{ij} defined by

$$C_{ij} = (1/e_i + 1/e_j - 1)^{-1}$$

Here the correlation $e_j + r_j = 1$ has been used.

Neglecting the motion of the gas, we can write the conductive heat flux through the gas between the substrate and the susceptor surface

$$J_{sf}^{cond} = \frac{\lambda(T_s - T_w)}{d_{sf} + 2\beta'l}$$

where the low Knudsen number heat transport is considered according to Equ.3.2 and d_{sf} is the distance between the susceptor surface and the wafer front side.

An analogous expression is used for the heat flux between the substrate and the cold walls

$$J_{bc}^{cond} = \frac{\lambda(T_w - T_c)}{d_{bc} + 2\beta'l}$$

with d_{bc} being the distance between the wafer back side and the cold reactor wall.

In the stationary case the heat balance at the wafer requires:

$$J_{sf}^{cond} - J_{bc}^{cond} + J_{sf}^{rad} - J_{bc}^{rad} = 0$$

From that the wafer temperature T_w can be calculated in the 1-D approximation

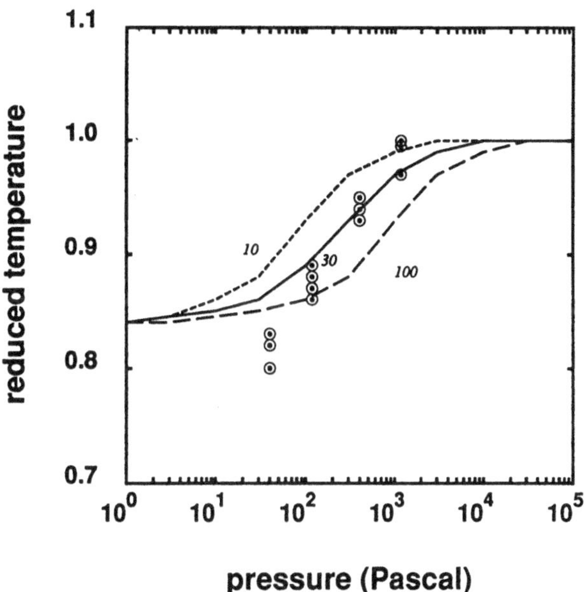

Figure 3.3: Comparison of the 1-D radiation model with experimental temperature values. Parameter is the value of the fitting parameter β'. The best agreement is found for $\beta'=30$

$$\frac{T_w - T_c}{T_s - T_c} = \frac{\frac{\lambda}{d_{sf}+2\beta'\ell T_s} + 4\sigma T_s^3(C_{sf} + C_{bc}) - \sigma C_{bc}(T_s^2 + T_c^2)(T_s + T_c)}{\frac{\lambda}{d_{sf}+2\beta'\ell T_s} + \frac{\lambda}{d_{bc}+2\beta'\ell T_c} + 4\sigma T_s^3(C_{sf} + C_{bc})} \tag{3.8}$$

where we have linearized the T_w^4 term as

$$T_w^4 = T_s^4 + 4T_s^3(T_w - T_s)$$

When we compare measured values of T_w for different pressures, we can vary the mean free path l over several orders of magnitude and derive a good fitting value for the empirical parameter β'. We have used the following values: d_{bc}=0.3m, d_{sf}=0.03mm, e_s=0.9 (graphite), e_w=0.2 (deposited tungsten, e_c=0.2 (polished steel), λ=0.2W/(mK) (hydrogen gas). For the mean free path ℓ the following expression was used

$$\ell = \frac{kT}{\sqrt{2}\pi a^2 p}$$

with the Boltzmann's constant $k = 1.38\cdot10^{-23}J/K$, the hydrogen molecule diameter $a = 2.6 \cdot 10^{-10}m$ and the pressure p in Pa.

Fig. 3.3 gives the reduced temperature $(T_w - T_c)/(T_s - T_c)$ as a function of pressure for different values of the fitting parameter β' together with measurements for the case T_c=300K and T_s=700K. The best agreement was obtained for $\beta' = 30$. Using Equ. 3.3 this would be consistent with an accomodation coefficient $\alpha = 0.1$.

Hasper et al. [3.4] have done similar 1-D temperature calculations using measured emissivities e_w and table data for the accommodation coefficient α from [3.5]. In their simulations they achieve very satisfactory agreement between calculated and measured temperature values without adjustable fitting parameters.

3.5.2   2-D model for the substrate temperature

The 1-D model was extended to a 2-D cylindrical symmetric geometry including also the radiative heat exchange J_{sc}^{rad} between the susceptor outside the silicon wafer and the cold walls and also the lateral heat loss S_L across the mechanical suspension between the susceptor and the cold walls as given in Fig.3.2. The parameter A of Equ. (3.7) was fitted to give a good agreement between the measured temperature profiles on suceptor and substrate at different pressure values. Moreover, the full convective and conductive heat transfer through the gas was considered according to the equation (2.4).

Fig. 3.4 gives the simulated temperature profiles for the best fit together with the experiment. In this simulation also the lateral heat flow both in the silicon substrate and the graphite susceptor was modelized using the heat conductivies of silicon ($40W/(mK)$) and graphite ($20W/(mK)$).

The heat inflow S_{in} from the lamp heaters was considered as a constant flux of heat and adjusted to yield the required temperature of 700 K on the susceptor edge. In order to get reasonable convergence rates the initial guess for the temperature was adapted to be roughly consistent with the heat inflow S_{in}.

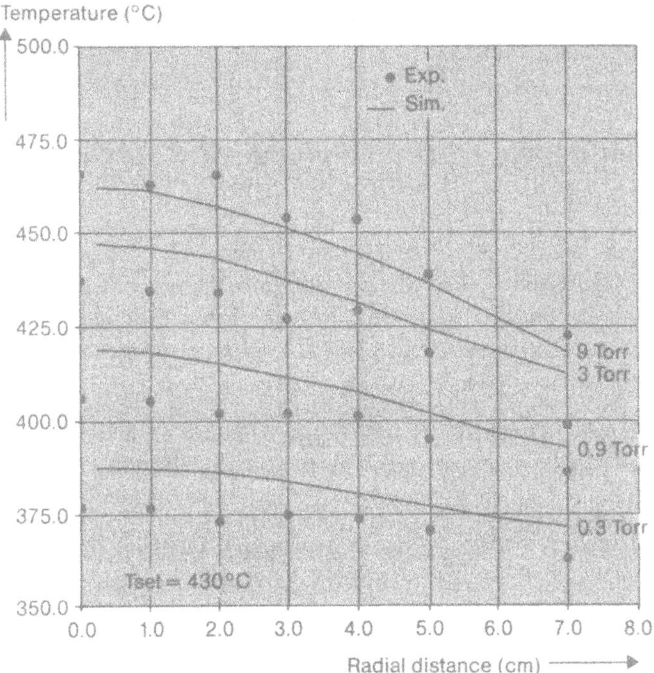

Figure 3.4: Radial temperature profiles on the wafer for different pressures. The fitting parameters β' and A in Equ. (3.2) and (3.7) were adjusted to fit the measured data.

It is obvious, that the discussed model contains a number of empirical fitting parameters and will not allow quantitative *a priori* predictions of temperature profiles in a new reactor. Nevertheless it has been advantageously used to support modifications of an existing reactor with the aim of improved temperature homogeneity.

For this purpose simulations were done for the case where the incoming heat flux from the lamps was screened from the center part of the susceptor, leading to an inhomogeneous heating arrangement. It has been proven that this arrangement was able to compensate in part for the lateral heat loss leading to a strongly improved homogeneity in temperature and hence in deposition rate.

Fig. 3.5 gives the calculated profiles of deposition rate for three different heating arrangements, where the incoming heat flux in the center part was screened by 1/3, 2/3 or fully, respectively. Also shown in the figure are the experimental values proving that the homogeneity was improved as much as a factor 2 in this case.

3.5.3   Modeling of the showerhead temperature

Fig. 3.6 shows a different reactor design, where the inlet is homogenized over a large area by means of a large plate located in some distance of the substrate which contains many small holes. From practical reasons it is important, that this showerhead plate is held at a moderate temperature to avoid deposition in the inlet holes.

The full 2-D radiation model according to equation (3.4) was used to calculate a consistent temperature profile on the showerhead plate, with

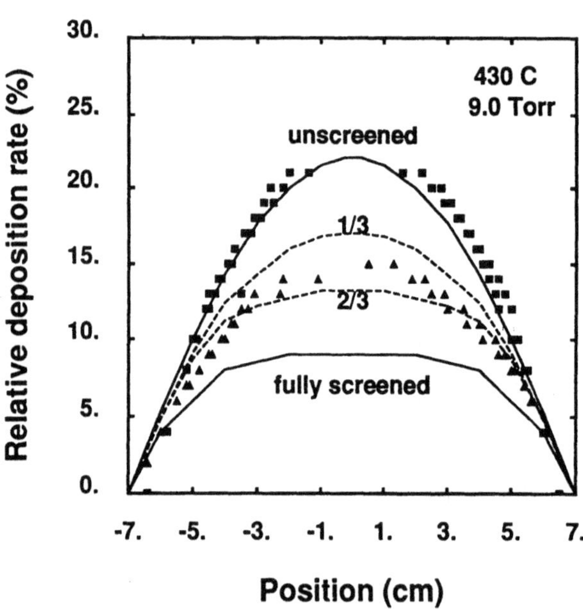

Figure 3.5: Deposition rate homogeneity for different heating arrangements. The symbols present measured data for the unscreened and the screened heating.

the correct consideration of the respective view factors F_{ij} between all surface elements.

Two different alternatives have been investigated. Fig. 3.7 shows temperature contour lines in the gas and in the shower plate for the case that the plate is thermally isolated from the cooled chamber walls and only exchanges heat by means of radiation and conduction through the gas. We see that the showerplate aquires a temperature about 390 K, which might be high enough to cause some trouble with deposition in the shower holes and hence require frequent cleaning cycles.

Fig. 3.8 shows a different possibility. Here the outside block holding the showerplate is also cooled and fixed to a temperature of 300 K. We see that the temperature on the showerplate is kept below 360 K for this arrangement.

In order to get quantitative results, the values for emissivity e_i of the susceptor, shower and reactor walls must be accurately known. Since these are strongly dependent on the surface quality (e.g. polished or rough) we have used different values for the emissivities to predict the range of

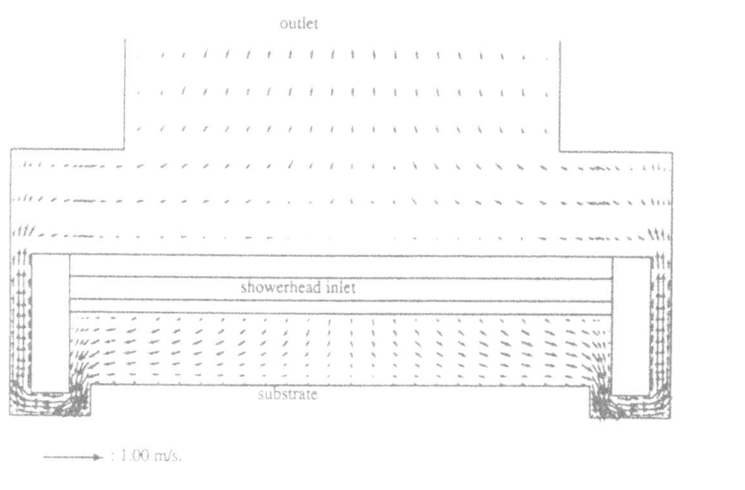

Figure 3.6: Tungsten CVD reactor with showerhead inlet. The temperature at the inlet plate must be kept low enough to avoid deposition in the inlet holes.

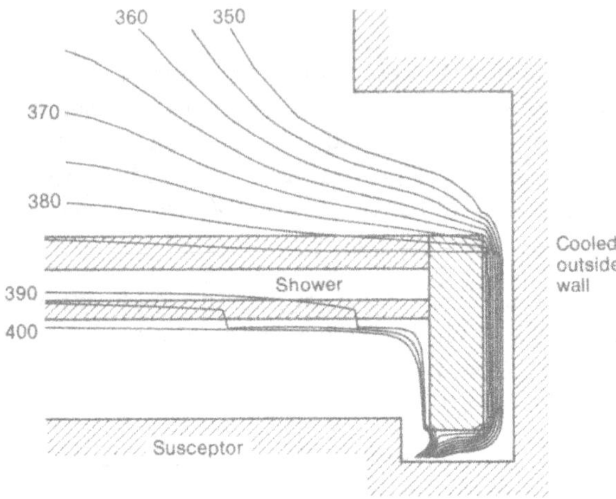

Figure 3.7: Temperature contours in the gas and in the showerhead for a thermally floating showerhead.

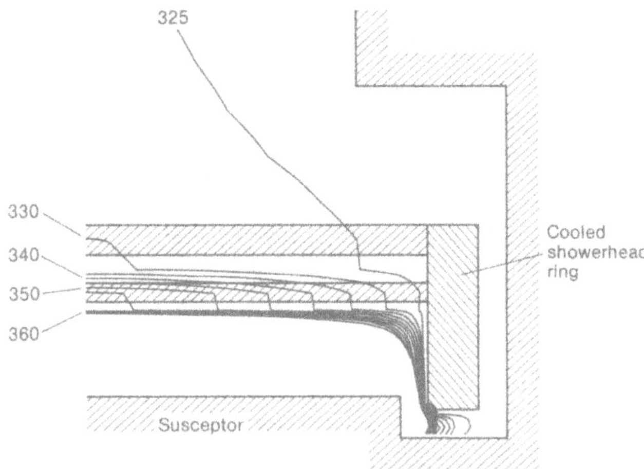

Figure 3.8: Temperature contours in the gas and in the showerhead for a cooled outside part of the showerhead.

temperature to be expected on the showerplate of such a reactor. Fig. 3.9 gives the respective temperature profiles for different choices of e_i' and shows that an uncertainty of about 15 K will result from changes in the emissivities.

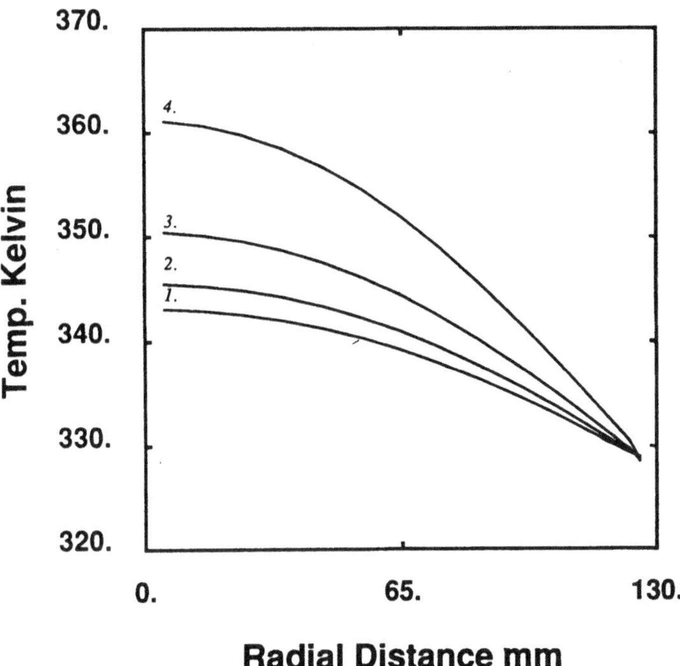

Figure 3.9: Temperature gradient across the inlet plate for the cooled showerhead arrangement. The different curves span the complete range of possible values for the emissivities.
1. no radiation
2. Ni 0.05 shower 0.15 W 0.06
3. Ni 0.17 shower 0.27 W 0.20
4. Ni 0.30 shower 0.40 Si(in place of W) 0.50

References Chapter 3

[3.1]   J.M. Lafferty, Editor (1965), Scientific Foundations
        of Vacuum Technique, Second Edition,
        John Wiley and Sons, Inc., New York.

[3.2]   Earle H. Kennard (1938), Kinetic Theory of Gases,
        McGraw-Hill Book Company, New York.

[3.3]   G.A.Bird (1988), "Direct simulation of gas flows at the
        molecular level", *Communications in Applied
        Numerical Methods*, Vol.4, pp.165-172

[3.4]   A. Hasper, J.E.J. Schmitz, F.Holleman and J.F. Verwey,
        (1992) "Heat transport in cold-wall single wafer reactors",
        J. Vac. Sci. & Techn. A, to be published

[3.5]   S.C. Saxena and R.K. Joshi, "Thermal accomodation and
        adsorption coefficients of gases", McGraw Hill/CINDAS
        Data Series on Material Properties, Vol II-1.

Chapter 4

Blanket tungsten deposition

4.1 Introduction

In recent years, the interest in the use of tungsten in multilevel metallization submicron IC technology has grown considerably [*e.g.* 4.1-4.7]. Tungsten is a suitable candidate for metallization purposes because it has a low bulk resistivity and a low contact resistance to $TiSi_2$, is chemically stable, has a high resistance to electromigration and has a coefficient of thermal expansion similar to that of silicon. Because of the poor step coverage, sputtering techniques are not suited for application in submicron devices and CVD techniques are required. For the filling of vias and contact holes, selective deposition processes, *i.e.* processes in which deposition takes place on metallic and silicon surfaces, but not on oxide, seem very attractive. However, until now efforts to develop commercially applicable selective tungsten LPCVD processes have not fully succeeded. Selective tungsten CVD processes are further discussed in chapter 5. Blanket tungsten deposition processes with subsequent backetching have been more successful. The filling of contact holes in a blanket mode requires deposition techniques capable of filling submicron, high aspect ratio holes with excellent conformality (high step coverage). Blanket tungsten LPCVD from tungstenhexafluoride (WF_6), using hydrogen as a reducing agent, according to the overall reaction:

$$WF_6(g) + 3H_2(g) \rightarrow W(s) + 6HF(g) \tag{4.1}$$

is now widely used for this purpose [4.8-4.10] and has been included in (pilot) production lines of most major IC manufacturers. The process was originally developed in hotwall multiple-wafer-in-tube LPCVD reac-

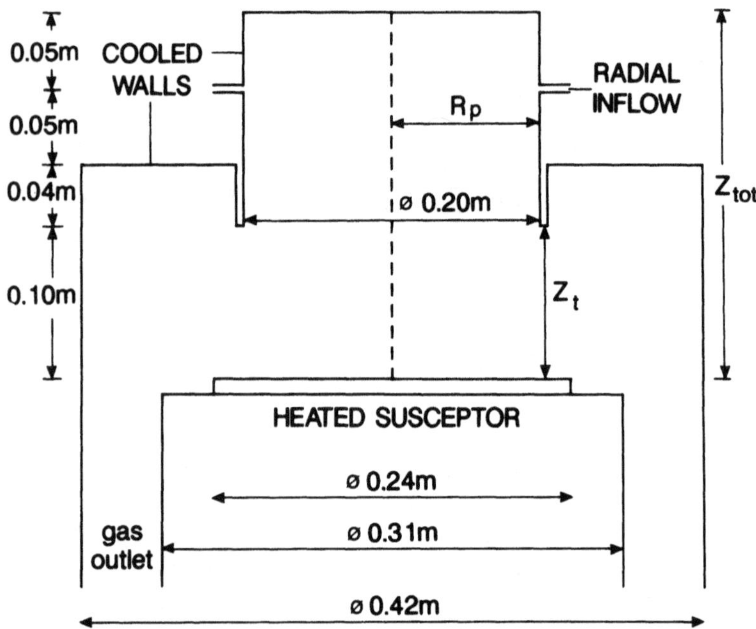

Figure 4.1: ASM cold-wall single-wafer LPCVD reactor

tors [4.11], but the use of coldwall single-wafer reactors is more common nowadays [4.12, 4.13] and offers several advantages over hotwall systems, such as low particle contamination, low WF_6 consumption, suppression of gas phase reactions, efficient removal of reaction by-products and easy automatic wafer handling.

The properties of the process and of the deposited tungsten films appear to depend strongly on the species concentrations, pressure and temperature distributions in the reactor and at the wafer surface. At sufficiently high WF_6 concentrations the growth rate is determined by the H_2 concentration and the temperature [4.12, 4.14], whereas at lower WF_6 concentrations it has been found to depend on the WF_6 concentration also [4.13, 4.15]. The step coverage appears to be improved by high WF_6 partial pressures, low temperatures and low H_2 partial pressures [4.16-4.20]. Encroachment (undesired W growth in contact windows just underneath SiO_2) has been found to decrease with decreasing temperature and WF_6 pressure [4.21]. From all these facts, it is clear, that it is important to have precise knowledge of the concentration and temperature distributions in tungsten LPCVD reactors. These distributions are highly determined by hydrodynamics and transport phenomena in the gas mixture in the reactor [4.22, 4.23]. Especially in single-wafer reactors with high deposition rates and cooled reactor walls, leading to a high rate of reactant consumption and reaction-product formation and to strong thermal diffusion effects, significant concentration and temperature gradients will be present. Detailed simulation models can be used to determine the species concentrations and temperature distributions in the gas mixture and at the wafer surface.

In this chapter, a mathematical CVD model based on the equations described in chapter 2 is used to study blanket tungsten LPCVD from H_2 and WF_6 in a coldwall single-wafer reactor. The main interests of the study are in (i) the growth rate and growth uniformity, (ii) the transition from kinetically-limited to transport-limited growth and (iii) the species concentration distributions at the wafer surface and their influence on step coverage. The studied reactor, which is illustrated in figure 4.1, has been developed by ASM International Inc. and has been designed for handling $0.20m$ diameter wafers. A small number of prototype reactors has been built and is being used as a research reactor for the development of tungsten LPCVD processes. The reactor is of the vertical impinging jet type and is cylinder symmetric. The walls of the stainless steel reactor chamber are water-cooled to room temperature. The wafer is placed on top of a $0.24m$ diameter graphite susceptor on top of a quartz dome. The susceptor is heated indirectly by means of a resistance heating element, which is fixed against the inside of the upper wall of the quartz dome. The parts of the dome outside the susceptor are protected from heating up by means of radiation shields. The gases are introduced radially into a $0.20m$ diameter gas injection tube, which is positioned perpendicularly above the susceptor surface, and leave the reactor through the outflow. In the simulations, it is assumed that the hydrodynamics are stationary and axisymmetric.

4.2 Chemistry model

For the common process conditions (600-750 K, 10^2-10^3 Pa), the LPCVD deposition rate of tungsten from H_2 and WF_6 appears to be fully determined by surface chemistry. A theoretical study of the reaction kinetics by Arora and Pollard [4.24] shows, that gas phase reactions are unimportant for these process conditions. The fact that the process can be run in a selective mode is another indication for the fact that the deposition is determined by heterogeneous reactions mainly. The overall reaction is given by *eq.* 4.1. For sufficiently high WF_6 concentrations, the deposition rate has been found to depend on the H_2 partial pressure and the temperature, being independent of the WF_6 partial pressure, according to

$$\mathcal{R}^s_{kin} = c_H [P_{WF_6}]^0 [P_{H_2}]^{\frac{1}{2}} \exp\left(\frac{-E_A}{RT}\right) \tag{4.2}$$

where $\mathcal{R}^s_{kin}$ is the reaction rate determined by the surface kinetics and the apparent activation energy $E_A \simeq$ 67-73 $kJ/mole$ [4.12, 4.14, 4.25-4.27]. At very low WF_6 concentrations however, the overall growth rate cannot remain independent of the WF_6 pressure, because of supply and mass transfer limitations. It is also possible, that the heterogeneous reaction rate changes from zero order to nonzero order in WF_6 at low concentrations. Indeed, at very low WF_6 partial pressures the growth rate has been found to decrease with decreasing WF_6 concentration. This has been ascribed to mass transfer limitations [4.13], and to a change in the reaction rate order in WF_6 from zero to 1/6 at low WF_6 concentrations [4.15]. In this chapter, it will be shown that the rate of the heterogeneous reaction remains zero order in WF_6 down to very low pressures.

We now assume, that the actual growth rate is described by a mechanism that considers the sequential processes of gas phase diffusion of reactants to the wafer surface and a heterogeneous reaction, as was done by Ulacia *et al.* [4.28]:

$$\frac{1}{\mathcal{R}^s_{eff}} = \frac{1}{\mathcal{R}^s_{kin}} + \frac{1}{\mathcal{J}^{max}_{WF_6}} + \frac{3}{\mathcal{J}^{max}_{H_2}} \tag{4.3}$$

Here, $\mathcal{R}^s_{eff}$ is the actual deposition rate, $\mathcal{R}^s_{kin}$ is the reaction rate determined by the heterogeneous reaction kinetics and $\mathcal{J}^{max}_i$ is the maximum diffusive mole flux of the ith species to the wafer surface. Through *eq.* 4.3, the slowest of the three mechanisms: (*i*) heterogeneous reaction at the wafer surface, (*ii*) diffusion of WF_6 to the wafer surface and (*iii*) diffusion of H_2 to the wafer surface determines the deposition rate. Usually, a large excess of H_2 is used. Also, the diffusion coefficient of H_2 in the gas mixture is much larger than that of WF_6. Therefore, $\mathcal{J}^{max}_{H_2}$ is usually much larger than $\mathcal{J}^{max}_{WF_6}$, and the deposition rate is determined either by the heterogeneous reaction kinetics, or by the supply of WF_6 to the wafer surface, or both.

Based on the extensive set of experimental growth rates obtained in a hotwall reactor published by Broadbent and Ramiller [4.14], $\mathcal{R}^s_{kin}$ was calculated from *eq.* 4.2, using

$$c_H = 1.7 \; mole \cdot Pa^{-\frac{1}{2}} \cdot m^{-2} \cdot s^{-1}$$
$$E_A = 69 \; kJ \cdot mole^{-1}. \tag{4.4}$$

with a statistical error of $\pm$ 0.5 $mole \cdot Pa^{-\frac{1}{2}} \cdot m^{-2} \cdot s^{-1}$ (95%) in the value of c_H at fixed E_A. The maximum mole flux $\mathcal{J}_i^{max}$ of the ith species to the wafer surface was taken as the maximum diffusive flux from the first discretization grid point to the wafer surface. The *actual* diffusive mole flux of the ith species from the first grid point to the wafer surface in the direction normal to this surface is given by

$$\mathcal{J}_i^{1 \to 0} = \frac{1}{m_i} \left(\rho \mathbb{D}_i \frac{\omega_i^1 - \omega_i^0}{\Delta} + \mathbb{D}_i^T \frac{\ln T^1 - \ln T^0}{\Delta} \right) \tag{4.5}$$

with m_i the species mole mass, $\mathbb{D}_i$ its effective ordinary diffusion coefficient, $\mathbb{D}_i^T$ its thermal diffusion coefficient, ω_i its mass fraction and T the temperature. The superscript ^{1} denotes the value in the first grid point next to the wafer surface, the superscript ^{0} denotes the value at the wafer surface, and Δ denotes the distance from the wafer surface to the first grid point. Now, the *maximum* diffusive mass flux to the wafer surface may be calculated from *eq.* 4.5, assuming that the wafer concentration ω_i^0 equals zero. As a result $\mathbb{D}_i^T$ will also be zero, since $\mathbb{D}_i^T \to 0$ for $\omega_i \to 0$. We now find for the maximum diffusive mole flux of the ith species from the first discretization point to the wafer surface

$$\mathcal{J}_i^{max} = \frac{\rho \mathbb{D}_i \omega_i^1}{m_i \Delta} \tag{4.6}$$

4.3 Simulation of blanket deposition

The transport equations described in chapter 2, together with the above chemistry model, were solved numerically in 2D axisymmetric form on a computational grid with 35 mesh points in radial and 35 mesh points in axial direction. Grid independence of the results was checked on 50×50 and 70×70 grids for some representative situations. As convergence criteria the error in the global mass balance for the total flow and for each of the the gas species, the residuals of the equations and the relative changes of the variables from one iteration to the next were used.

In figure 4.2 some examples of simulation results are shown. Figures 4.2a and 4.2b show calculated streamlines and isotherms for a total pressure

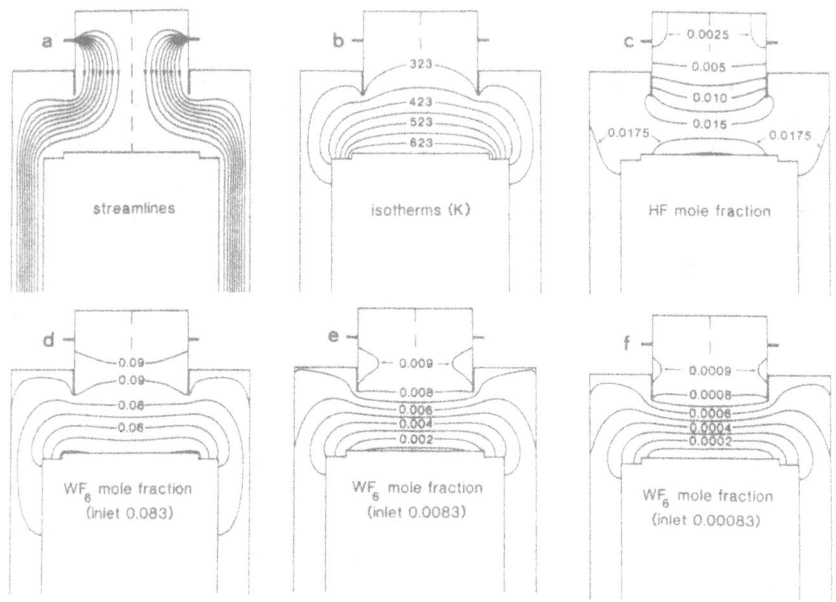

Figure 4.2: Model predictions for 133 Pa total pressure, 673 K wafer temperature, 1000 $sccm$ H_2 and 200 $sccm$ $Ar + WF_6$. (a-d $= 100$ $sccm$ WF_6; e $= 10$ $sccm$ WF_6, f $= 1$ $sccm$ WF_6)

of 133 Pa, a wafer temperature of 673 K and an inlet flow of 1000 $sccm$ H_2, 100 $sccm$ WF_6 and 100 $sccm$ argon. At these process conditions the flow is not disturbed by buoyancy effects and no flow recirculations are observed, even though the reactor walls are cooled and the heated susceptor is facing upward. It can also be seen, that there is no thin thermal boundary layer above the susceptor. Instead, the heated gas region extends over a large part of the reactor volume. The growth rate is fully determined by surface reaction kinetics and is very uniform, as is shown in figure 4.3. Due to the 460 $\mathring{A}ng/min$ deposition on the 0.20m diameter wafer, 3.6 $sccm$ WF_6 and 10.8 $sccm$ H_2 are consumed and 21.6 $sccm$ HF is produced. In figures 4.2c and 4.2d the resulting HF and WF_6 mole fractions in the reactor are shown. Although only 3.6% of the incoming WF_6 is consumed, the WF_6 concentration at the wafer surface is less than 50% of the inlet concentration. This is mainly due to the strong thermal

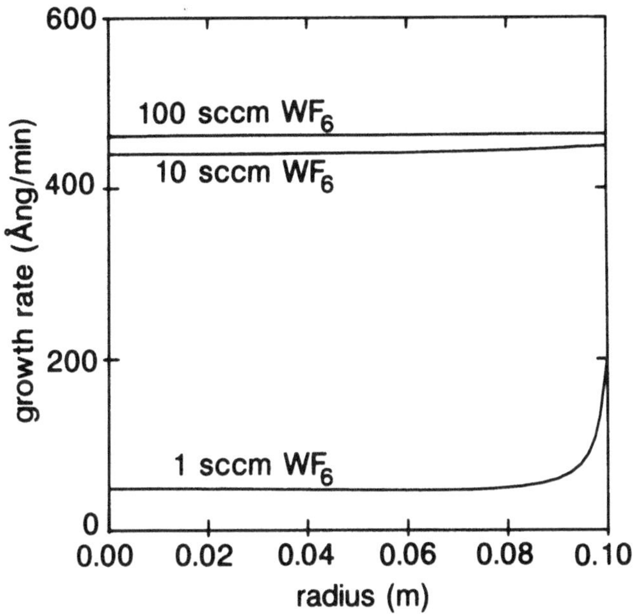

Figure 4.3: Predicted growth rates for 133 Pa total pressure, 673 K wafer temperature, 1000 $sccm$ H_2 and 200 $sccm$ $Ar + WF_6$

diffusion effects, causing the heavy WF_6 molecules to move away from the hot susceptor. Furthermore, a concentration gradient between inlet and wafer surface is required in order to have diffusive transport of WF_6 to the wafer surface. In figure 4.2e the incoming WF_6 flow is reduced to 10 $sccm$, whereas the Ar flow is increased to 190 $sccm$, keeping all other parameters fixed as in figure 4.2a-d. In this case the growth rate is just beginning to be limited by WF_6 transport and has slightly decreased to 420 $Ång/min$ (figure 4.3). Although the WF_6 consumption is only 33%, the WF_6 concentration at the wafer surface is less than 10% of the inlet concentration. In figure 4.2f the WF_6 inlet flow is further reduced to 1 $sccm$. Now, the growth rate is strongly limited by WF_6 transport and highly non-uniform (figure 4.3). The WF_6 concentration at the wafer surface is zero and the WF_6 consumption is 45%.

Figure 4.4 shows simulation results for the deposition rate, the ratio of the WF_6 pressure at the wafer surface and in the reactor inlet, and the fraction

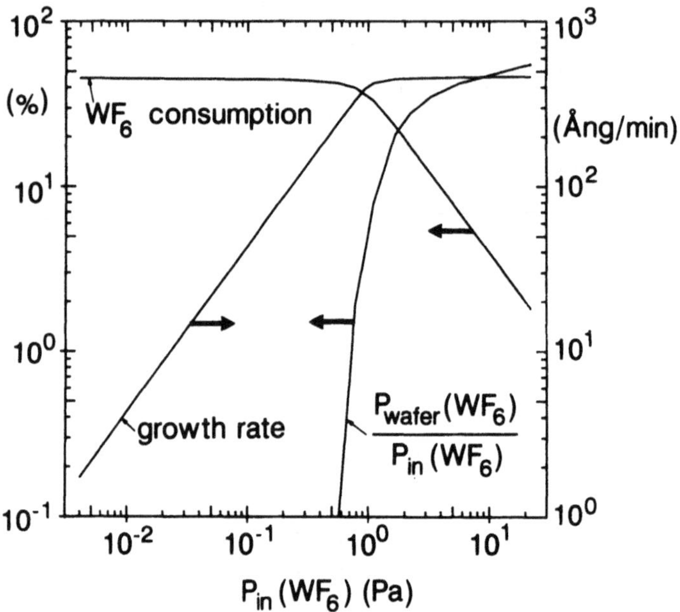

Figure 4.4: Predicted growth rate, WF_6 surface pressure and WF_6 consumption as a function of the WF_6 inlet pressure, for 133 Pa total pressure, 111 Pa H_2 pressure, 673 K wafer temperature and 1.2 slm total flow

of consumed (=deposited) WF_6, as a function of the WF_6 inlet pressure for the above process conditions. For WF_6 inlet pressures $P_{in}(WF_6)$ above a certain value P_{crit} the growth rate is kinetically limited and independent of $P_{in}(WF_6)$, whereas for lower $P_{in}(WF_6)$ the growth rate becomes transport limited and decreases linearly with decreasing $P_{in}(WF_6)$. The WF_6 consumption is inversely proportional to $P_{in}(WF_6)$ for $P_{in}(WF_6) > P_{crit}$, whereas for lower $P_{in}(WF_6)$ the consumption reaches a maximum, which, for the present conditions, is 45% of the WF_6 inflow. The WF_6 pressure at the wafer surface is substantially lower than the WF_6 inlet pressure for all process conditions studied. For $P_{in}(WF_6) > P_{crit}$ however, the surface pressure is of the same order of magnitude as the inlet pressure, whereas for $P_{in}(WF_6) < P_{crit}$ the WF_6 wafer pressure quickly drops to zero.

The above illustrates the influence of the WF_6 inlet pressure on the transition from kinetically limited to transport-limited growth, on the WF_6 concentration on the wafer surface, on the WF_6 consumption and on the growth rate and uniformity.

4.4 Approximations for multicomponent diffusion

In the previous section we have used the rigorous treatment of multicomponent ordinary diffusion described in section 2.4.2 (Stefan-Maxwell equations, *eqs.* 2.11-2.16) and the exact formulation for multicomponent thermal diffusion described in sections 2.4.3 (*eq.* 2.19) and 2.7.4 (*eqs.* 2.89-2.96). In section 2.4.2 (*eqs.* 2.17-2.18) we have described an approximation for modeling multicomponent ordinary diffusion which (*i*) requires less computational effort and (*ii*) leads to expressions which can be implemented easily as a standard gradient diffusion term in the general transport equation 2.97. In section 2.7.4 (*eqs.* 2.75-2.86, 2.88) we have described an approximate method for calculating multicomponent thermal diffusion coefficients, which requires much less computational effort than the exact formulation. In this section we will study the accuracy of both approximations for the modeling of tungsten LPCVD from hydrogen and tungstenhexafluoride.

For this process the gas mixture in the reactor consists of WF_6, H_2, Ar and HF. We compare the six approaches in table 4.1. It can be seen that, for the present model implementations, the approximation for thermal diffusion leads to a 34% reduction of the required cpu time per iteration compared to the exact formulation, whereas the approximation for ordinary diffusion leads to another 20% reduction. The inclusion of the Dufour effect requires relatively little additional cpu time. The total number of iterations required to obtain a converged solution is approximately the same for all the different approaches.

We now compare the accuracy of these six approaches for two different process conditions:
(*i*) A process in the kinetically-limited growth regime, with 133 Pa total pressure, 673 K wafer temperature, 1000 $sccm$ hydrogen, 100 $sccm$ argon and 100 $sccm$ tungstenhexafluoride flow. In this case, the growth rate is almost unaffected by transport phenomena in the gas mixture, but the species concentrations at the wafer surface are determined by convection and (thermal) diffusion phenomena.

Table 4.1: Approximations for multi-component (thermal) diffusion

no.	ordinary diffusion	thermal diffusion	Dufour effect	Normalized CPU time per iteration	WF_6 partial pressure (Pa) at wafer surface for kinetically limited process	Growth rate (Ång/min) for transport limited process
I	exact	exact	included	1.00	5.26	47.8
II	exact	exact	not incl.	0.97	5.27 (+ 0%)	47.8 (+ 0%)
III	exact	approx.	included	0.66	6.81 (+29%)	49.7 (+ 4%)
IV	exact	not incl.	not incl.	0.61	10.1 (+92%)	60.9 (+27%)
V	approx.	exact	included	0.80	6.69 (+27%)	52.6 (+10%)
VI	approx.	approx.	not incl.	0.46	7.95 (+51%)	54.3 (+14%)

(*ii*) A process in the diffusion-limited growth regime, with 133 *Pa* total pressure, 673 *K* wafer temperature, 1000 *sccm* hydrogen, 199 *sccm* argon and 1 *sccm* tungstenhexafluoride flow. In this case, the growth rate is determined by the transport phenomena in the reactor.

The results of these comparisons are again presented in table 4.1. The influence of the Dufour effect is negligible (compare I *vs* II), but the effect of thermal diffusion is very important (compare I *vs* IV). For this process, the approximation for multi-component thermal diffusion is not very accurate (compare I *vs* III). Also, the use of the approximation for multi-component ordinary diffusion instead of the full Stefan-Maxwell equations leads to rather large errors (compare I *vs* V). Therefore, for the accurate modeling of tungsten LPCVD from WF_6 and H_2, the exact formulations for multicomponent ordinary and thermal diffusion should be used.

4.5　Experimental validation of growth rate simulations

To validate the simulation model, predicted growth rates and uniformities were compared to experimental results [1]. Furthermore, the calculated WF_6 surface concentrations were used as input for a step coverage model, and the predicted step coverages from these combined two models were

---

[1] The experiments described in sections 4.5 and 4.6 were performed by Dr. A. Hasper and J. Holleman of the MESA Institute of Twente University (The Netherlands). See also references 4.19, 4.20, 4.22 and 4.23.

compared to experimental results. The latter is further described in section 4.6. Special attention is given to process conditions with low WF_6 flows, since, for economic reasons, it is interesting to use as little WF_6 as possible. At these conditions, both the growthrate and the step coverage are strongly dependent on the WF_6 surface concentration, thus offering a sensitive way to test the model. At the same time more insight is gained in the processes influencing the deposition at low WF_6 concentrations.

4.5.1 Experimental method

In the experiments, 3 *in.* (0.076m) diameter p type 10 Ω *cm* (100) wafers were used, which were placed in the center of an 8 *in.* silicondioxide coated carrier wafer. Immediately before loading the wafer into the reactor a HF (1:100) dip of 30 seconds was applied. The purity of the gas sources (WF_6, Ar, H_2) employed was 99.999%, according to manufacturers specifications. Very low WF_6 flows could be realized by putting a needle valve in series with the WF_6 massflow controller. Thus it was possible to adjust the WF_6 flow down to 0.4 *sccm* with an accuracy of $\pm$ 0.05 *sccm*. The silicon wafer temperature was measured by means of a dual wavelength pyrometer through a quartz window in the upper wall of the reactor. In this way, the wafer temperature (which may deviate substantially from the susceptor temperature in coldwall LPCVD reactors as was shown in chapter 3) was known with an accuracy of $\pm$ 5 K at the beginning of the deposition process. During the deposition process, the susceptor temperature as measured by a series of thermocouples was kept constant.

In the growth rate experiments, deposition took place on the 3 *in.* wafer surface only and was stopped immediately at the moment selectivity was lost and deposition at the carrier wafer and susceptor was initiated. Layer thicknesses were determined by measuring the weight increase of the wafer and by using a profilometer. The measured layer thickness was corrected for the thickness of the initial, silicon reduced tungsten layer. By determining the weight increase, the wafer averaged growth rate could be determined with an accuracy of $\pm$ 20 $\mathring{A}ng/min$. The profilometer measures the local thickness, thus allowing the determination of growth rate uniformities. Profilometer measurements however suffer from inaccuracies due to surface roughnesses, which may be 10% of the total layer thickness for H_2 reduced tungsten films.

For the step coverage experiments, rectangular trenches with depths ranging from 5 to 10 μm and widths ranging from 1 to 5μm were etched into the silicon wafer surface. In these experiments, deposition took place on the 3 *in.* silicon wafer as well as on the 8 *in.* oxide coated carrier wafer and on the heated susceptor. The step coverage was determined by Scanning Electron Microscope observations with an accuracy of $\pm$ 5%.

4.5.2　Experimental results and comparison with model simulations

In figure 4.5 experimental results for the wafer averaged growth rate as a function of the WF_6 inlet pressure are shown. In this series of experiments the wafer temperature was 673 K the total pressure was 133 Pa, and the H_2 flow was 1000 *sccm*. The WF_6 flow was varied from 0.4 to 200 *sccm*, thus varying the WF_6 inlet pressure by almost three orders of magnitude from 0.044 to 22 Pa. Argon was used to keep the total flow fixed at 1200 *sccm*. Also shown are results of model simulations for the same process parameters. The model fairly accurately predicts the WF_6 independent growth rate for large WF_6 inlet pressures, the linear dependence of the growth rate on the WF_6 inlet pressure for low inlet pressures and the transition point P_{crit} between these two regions. It should be noted, that the model predictions in figure 4.5 were obtained without the introduction of any fitting parameter. The model assumes a heterogeneous reaction rate which is zero order in WF_6 for all WF_6 concentrations. Thus, the decreasing growth rates for low WF_6 concentrations predicted by the model are caused by mass transfer limitations. Because of the good agreement between model predictions and experimental observations, it may be concluded that decreasing growth rates at low WF_6 inlet concentrations are indeed caused by mass transfer limitations rather than a change in the reaction kinetics. More arguments supporting this conclusion will be discussed below.

When comparing model predictions and experimental results in detail, it can be seen that the model systematically overpredicts the growth rate at low WF_6 concentrations. This may possibly be explained by relatively small errors in the transport properties, especially in the (thermal) diffusion coefficients as obtained from kinetic theory (section 2.7). When *e.g.* the binary ordinary diffusion coefficient for H_2 and WF_6 was reduced by 40% compared to the theoretical value, keeping all other parameters fixed, a very good match between model predictions and experimental results was obtained. An 40% error in the predicted diffusion coefficients for a gaspair like H_2 and WF_6 at high temperatures is not very unlikely, especially since the Lennard-Jones force parameters for WF_6 had to be es-

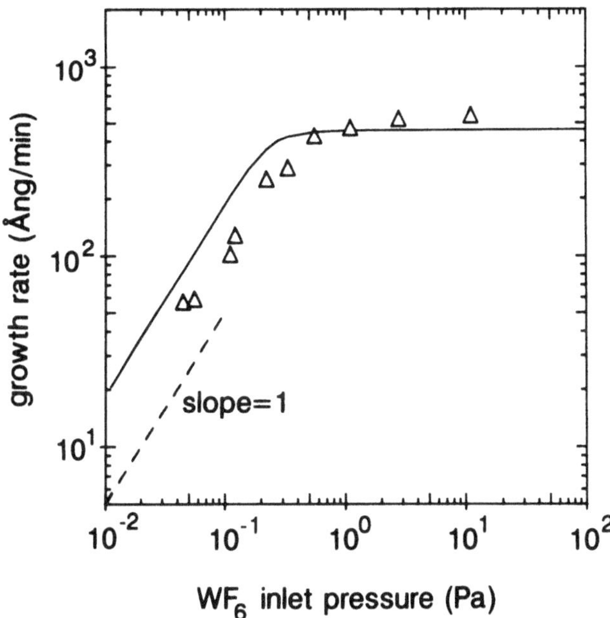

Figure 4.5: Model predictions and experimental data for the 3 *in.* wafer averaged growth rate wafer as a function of the WF_6 inlet pressure (133 Pa total pressure, 111 Pa hydrogen inlet pressure, 673 K wafer temperature, 1200 *sccm* total flow)

timated from critical parameters (table 2.4). Therefore, a corrected binary ordinary diffusion coefficient for H_2 and WF_6 of 0.6 times the theoretical value has been used in all further model simulations.

In figure 4.6, the experimental results from figure 4.5 are shown again, together with simulation results, using the corrected H_2-WF_6 diffusion coefficient. Also shown are comparisons between experimental and simulated growth rates for series of experiments at different wafer temperature and total pressure. With the corrected diffusion coefficient, the model accurately predicts the growth rate as a function of the WF_6 inlet pressure for all cases considered. Both the experiments and the model simulations show, that the growth rate at low WF_6 concentrations is linearly proportional to the WF_6 inlet pressure, inversely proportional to the total pressure and independent of the wafer temperature (at fixed total flow and

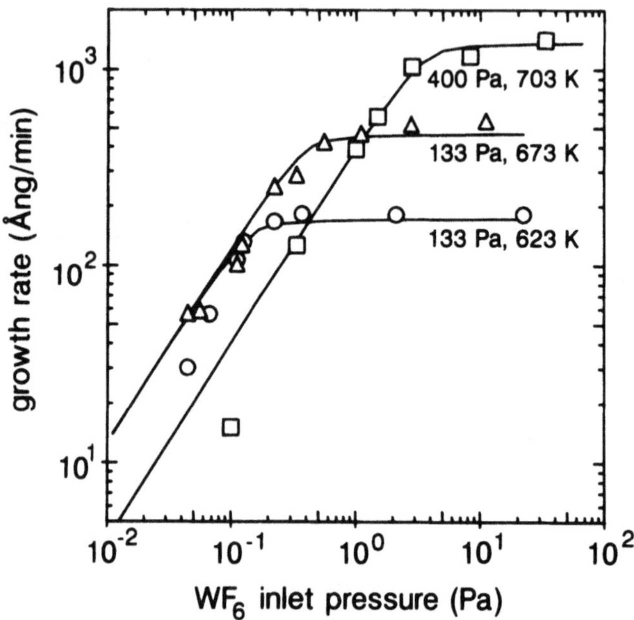

Figure 4.6: Model predictions and experimental data for the wafer averaged growth rate as a function of the WF_6 inlet pressure at different wafer temperatures and total pressures. (1000 *sccm* H_2, 200 *sccm* $Ar + WF_6$)

reactor geometry). This again is an indication for the fact, that decreasing growth rates at low WF_6 inlet concentrations are caused by mass transfer limitations rather than kinetical effects. Diffusion coefficients vary relatively little with temperature, causing the mass-transfer-limited growth rate to be almost independent of temperature. On the other hand, diffusion coefficients are inversely proportional to the total pressure, causing the mentioned effect of total pressure on the transfer-limited growth rate at fixed WF_6 pressure.

When (at fixed wafer temperature, total pressure and inlet species partial pressures) the total flow is increased, the convective transport of WF_6 to the wafer surface is increased, so WF_6 will be less depleted and its concentration at the wafer surface will increase. Thus it is expected that the mass-transfer-limited growth rate increases with increasing total flow. This was confirmed by two experiments at a total pressure of 133 Pa,

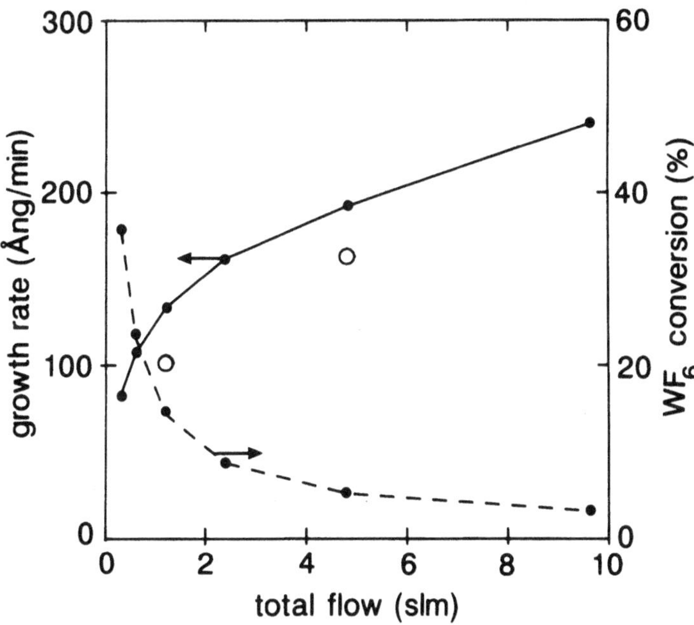

Figure 4.7: Model predictions and experimental data for the wafer averaged growth rate as a function of the total flow. The dashed curve shows the calculated WF_6 conversion. (0.11 Pa WF_6 and 111 Pa H_2 inlet pressure, 133 Pa total pressure, 673 K wafer temperature)

WF_6 and H_2 inlet partial pressures of 0.11 Pa and 111 Pa, and a wafer temperature of 673 K, and total flows of 1200 and 4800 $sccm$ respectively. For these process conditions, *eqs.* 4.2 and 4.4 predict a surface kinetics determined growth rate of 460 $\mathring{A}ng/min$. The observed transport limited growth rates for these two experiments were 101 and 163 $\mathring{A}ng/min$ respectively. In figure 4.7 these results are shown, together with model simulations for total flows between 300 and 9600 $sccm$. As expected, an increased total flow leads to an increased mass-transfer-limited growth rate. However, the effect of a large increase in total flow is relatively small, indicating that transport of WF_6 from reactor inlet to wafer surface is mainly due to diffusion rather than convection. It is also clear that a low WF_6 consumption does not guarantee the exclusion of mass transfer limitations. Even at a total flow of 10 slm and a WF_6 conversion of only

3% the growth rate is highly reduced by mass transfer limitations.

In figure 4.8 experimental and modeling results for the growth rate as a function of radius on a 3 *in.* wafer are shown for a total pressure of 133 *Pa*, a wafer temperature of 673 *K* a H_2 flow of 1000 *sccm*, a WF_6 flow of 1.3 *sccm* and an *Ar* flow of 198.7 *sccm*. The kinetically determined growth rate according to *eqs.* 4.2 + 4.4 is again 460 *Äng/min*. Both model predictions and experiments show a highly reduced, strongly non-uniform, mass-transfer-limited deposition.

4.6   Step coverage model and validation

When filling high aspect ratio contact holes using blanket tungsten deposition it is important that the step coverage is high and uniform over the entire wafer surface. In figure 4.9 a pore with initial diameter W_0 and depth L_0 is illustrated. The step coverage is defined as the ratio $d_{\frac{1}{2}}/(\frac{1}{2}W_0)$ in percents, with $d_{\frac{1}{2}}$ the layer thickness halfway the pore depth at the moment of pore closure. A low step coverage will lead to undesirable void formation when filling contact holes. In this section the use of a CVD reactor model as an aid in the prediction of step coverages will be discussed.

Since the dimensions of vias and contact holes in IC's are small compared to the mean free path length of the molecules at typical LPCVD conditions, transport of reactants in the pore is due to Knudsen diffusion. The Knudsen diffusion coefficient for the ith species in a high depth-to-width aspect ratio cylindrical pore or rectangular trench is given by [4.29, 4.30]

$$\mathbb{D}_i^K(\mathcal{W}) = \frac{2}{3}\mathcal{W}(\frac{8RT}{\pi m_i})^{\frac{1}{2}} \tag{4.7}$$

where $\mathcal{W}$ is the radius $\frac{1}{2}w$ of the pore or the width w of the trench. We can now find an expression for the step coverage as a function of the pore or trench dimensions, the deposition rate, the WF_6 concentration at the pore mouth and its Knudsen diffusion coefficient. Such an analysis was first presented by McConica and Churchill [4.16]. Consider a high aspect ratio pore (trench) as in figure 4.10. From an overall mass balance, equating the diffusive transport of WF_6 through the pore (trench) mouth to the amount of WF_6 deposited on its walls, we find that tungsten growth at a certain depth $z = z_C$ will stop when the pore radius (trench width) $\mathcal{W}$ has diminished to a certain critical value $\mathcal{W}_C$, with

$$\frac{\pi \mathcal{W}_C^2 \mathbb{D}_{WF_6}^K(\mathcal{W}_C)c_{WF_6}^0}{z_c} = 2\pi \mathcal{W}_C z_C \mathcal{R}_{kin}^s \tag{4.8}$$

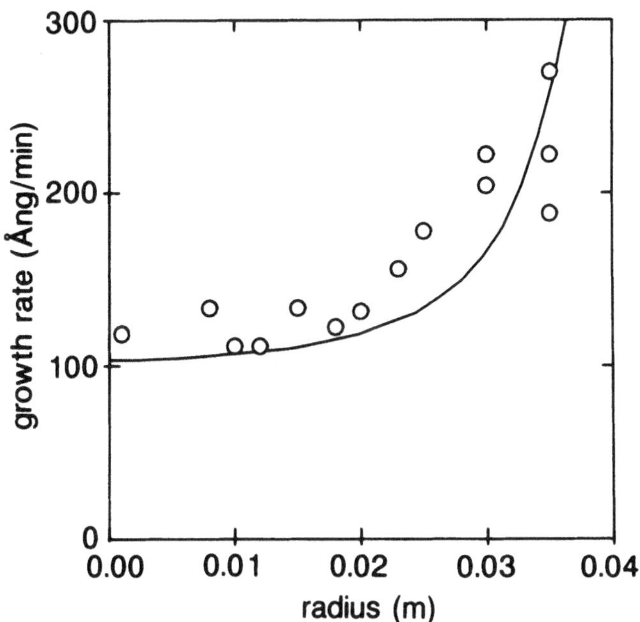

Figure 4.8: Model predictions and experimental data for the growth rate as a function of radius. (133 Pa total pressure, 673 K wafer temperature, 1000 $sccm$ H_2, 1.3 $sccm$ WF_6 and 198.7 Ar)

for a pore, and

$$\frac{\mathcal{W}_C \mathbb{D}^K_{WF_6}(\mathcal{W}_C)c^0_{WF_6}}{z_c} = 2z_C \mathcal{R}^s_{kin} \tag{4.9}$$

for a trench. Here $c^0_{WF_6}$ is the WF_6 concentration at the pore mouth. So, for trenches and pores, we have

$$\mathcal{W}_C = \frac{2z_C \mathcal{R}^s_{kin}}{\mathbb{D}^K_{WF_6}(\mathcal{W}_C)c^0_{WF_6}} \tag{4.10}$$

Since $\mathbb{D}^K_i(\mathcal{W}_C) = (\mathcal{W}_C/\mathcal{W}_0)\mathbb{D}^K_i(\mathcal{W}_0)$, we find that tungsten growth at $z_C = \frac{1}{2}L_0$ will stop when

$$\mathcal{W}_C^2 = \frac{L_0^2 \mathcal{R}^s_{kin} \mathcal{W}_0}{2\mathbb{D}^K_{WF_6}(\mathcal{W}_0)c^0_{WF_6}} \tag{4.11}$$

We now introduce the dimensionless number Φ, which is often referred to

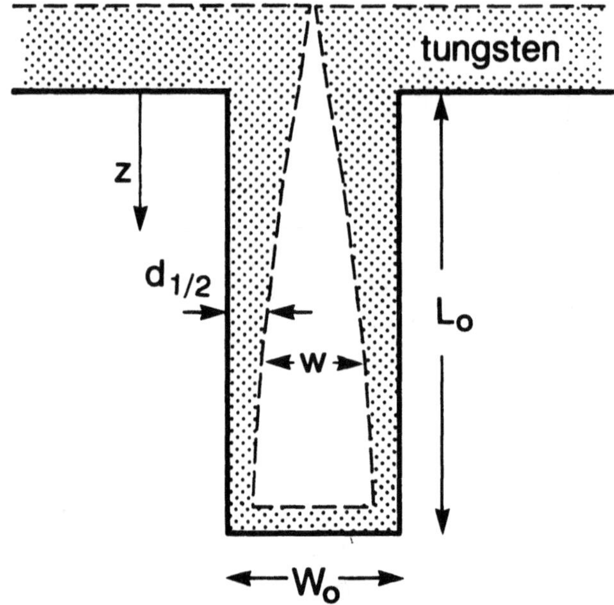

Figure 4.9: A high aspect ratio pore with deposited tungsten

as the step coverage modulus,

$$\Phi = \frac{L_0^2 \mathcal{R}_{kin}^s}{\mathcal{W}_0 \mathbb{D}_{WF_6}^K(\mathcal{W}_0) c_{WF_6}^0} \tag{4.12}$$

leading to the following expression for the step coverage in a cylindrical hole or rectangular trench

$$step\ coverage = (1 - \frac{\mathcal{W}_C}{\mathcal{W}_0})100\% = (1 - \sqrt{(\frac{1}{2}\Phi.)})100\% \tag{4.13}$$

¿From *eqs.* 4.7, 4.12 and 4.13 it can be seen, that the step coverage modulus Φ and the step coverage do not depend on the pore or trench dimensions, but on the (initial) aspect ratio $L_0/\mathcal{W}_0$ only. For a trench, Φ is four times smaller than for a cylindrical pore of equal initial depth and width. Figure 4.10 shows the step coverage in trenches and contact holes as a function of the step coverage modulus Φ according to *eq.* 4.13. A high step coverage is obtained when Φ is low. A 100 % step coverage can

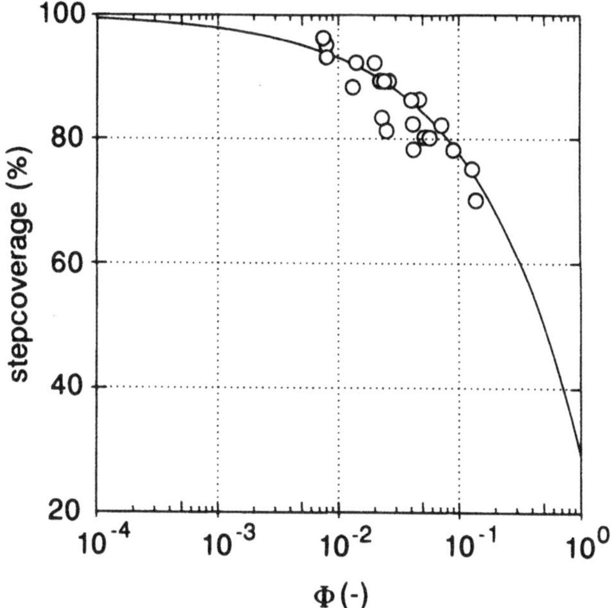

Figure 4.10: Theoretical and experimental step coverage as a function of the step coverage modulus

never be obtained.

It is now clear that the step coverage is a function of the WF_6 concentration at the wafer surface. At fixed aspect ratio, wafer temperature and deposition rate, Φ is inversely proportional to the WF_6 concentration at the pore mouth and a high step coverage is obtained for high WF_6 concentrations. Thus, in order to predict the step coverage, the WF_6 surface concentration must be known. Therefore, we have used the CVD reactor simulation model to calculate the WF_6 surface concentration. The calculated WF_6 surface concentrations were then used to calculate the step coverage modulus from *eq.* 4.12 and the theoretical step coverage from *eq.* 4.13. This was done for trenches of varying dimensions (5-10 μm depth, 1-5 μm width, aspect-ratio 2-7) for a series of experiments at varying process conditions (133-1064 Pa total pressure, 0.24-180 Pa WF_6 inlet pressure, 648-713 K wafer temperature, 0.3-5 slm total flowrate). For these process conditions and trench dimensions, Φ varied from 0.008

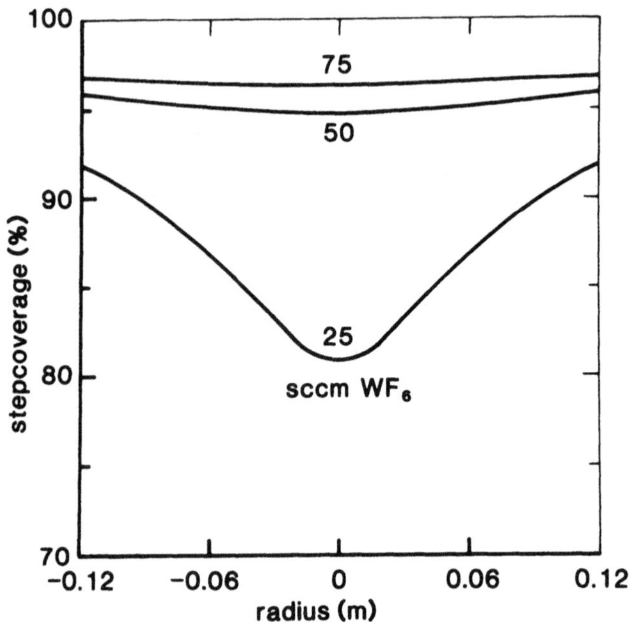

Figure 4.11: Predicted step coverage for varying WF_6 flows. (1064 *Pa* total pressure, 693 *K* wafer temperature, 200 *sccm* H_2, 100 *sccm* WF_6 + *Ar*)

to 0.14 and the experimental step coverage varied from 95 % to 70 %. A good agreement was found between theoretical step coverages predicted from *eqs.* 4.12-4.13 and experimental values obtained from SEM cross sections, see figure 4.10. This may serve as a confirmation of the accuracy of the WF_6 surface concentrations predicted by the reactor simulation model.

4.7   Process and reactor optimization

The ideal blanket tungsten contact fill process should combine a high and uniform growth rate and a high and uniform step coverage with a low use of (expensive) WF_6 gas. On the other hand, the allowable process window is limited by several constraints. The most important constraint is

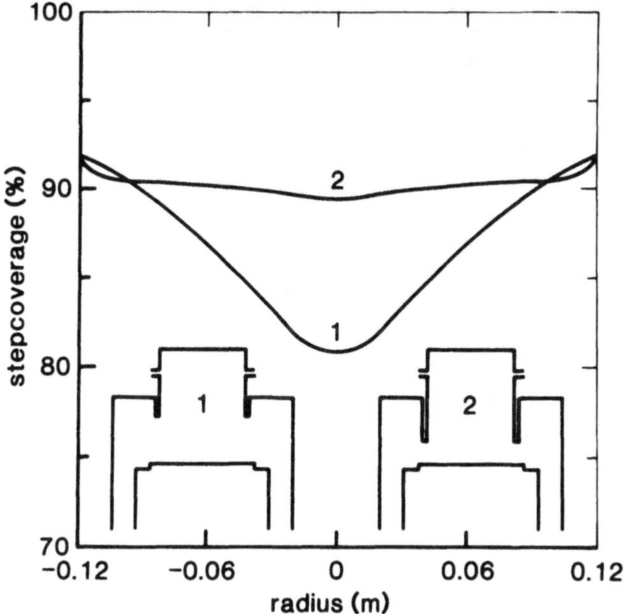

Figure 4.12: Predicted step coverage for different reactor geometries. (1064 *Pa* total pressure, 693 *K* wafer temperature, 200 *sccm* H_2, 25 *sccm* WF_6, 75 *sccm Ar*)

the maximum process temperature, which, with regard to multilevel metallization, is limited to *circa* 700 *K*. Furthermore, for the studied reactor, the maximum total pressure is limited to 10 *Torr* (1330 *Pa*).

We have used the combination of the reactor simulation model and the step coverage model in the design of such a process. Extensive series of model simulations were performed to optimize the process conditions, considering the above constraints. As an example of such optimizations studies, figure 4.11 shows the predicted step coverage in a 5:1 trench, as a function of the location of the trench on the wafer surface, for varying WF_6 flows. A relatively low and highly non-uniform step coverage is obtained for a WF_6 flow of 25 *sccm*. When this flow is increased to 50 *sccm*, the step coverage is highly improved. A further increase to 75 *sccm* does not further improve the process significantly. Similarly, figure 4.12 shows the influence of a small geometric optimization in the reactor configuration,

Figure 4.13: SEM cross section picture of a 9×1.5 μm trench filled with tungsten by means of the optimized process

which may highly improve the step coverage.

Thus, computer simulations have lead to the following optimized process conditions: 8 *Torr* (1064 *Pa*) total pressure, 693 *K* wafer temperature, 200 *sccm* H_2 + 50 *sccm* WF_6 + 50 *sccm* *Ar* flow. The predicted process characteristics are: 1450 $\mathring{A}ng/min$ growthrate ($\pm$ 1% over 8 *in.* wafer) and 95.6 % stepcoverage ($\pm$ 0.3 % over 8 *in.* wafer) in 5:1 (depth-to-width) trenches or 5:4 (depth-to-diameter) cylindrical holes. Such a process would for instance allow the void-free blanket filling of 0.6 μm diameter vias with a 0.75 μm depth, processing an 8 *in.* wafer within 2 minutes of deposition time. Experimentally, the optimized process conditions lead to a growth rate of 1430 $\mathring{A}ng/min$ and a step coverage $>$93 % in 9×1.5 μm trenches (both uniform within the experimental error over a 6 *in.* wafer). Figure 4.13 shows a SEM cross section of such a trench, filled with tungsten by means of the above process, illustrating the extraordinary good step coverage obtained with this process.

References Chapter 4

[4.1] R.S. Blewer (*ed.*) (1986) "Tungsten and Other Refractory Metals for VLSI Applications", The Materials Research Society, Pittsburgh, USA

[4.2] E. Broadbent (*ed.*) (1987) "Tungsten and Other Refractory Metals for VLSI Applications II", The Materials Research Society, Pittsburgh, USA

[4.3] V. Wells (*ed.*) (1988) "Tungsten and Other Refractory Metals for VLSI Applications III", The Materials Research Society, Pittsburgh, USA

[4.4] R.S. Blewer and C.M. McConica (*eds.*) (1989) "Tungsten and Other Refractory Metals for VLSI Applications IV", The Materials Research Society, USA

[4.5] S.S. Wong and S. Furukawa (*eds.*) (1990) "Tungsten and Other Advanced Metals for VLSI/ULSI Applications V", The Materials Research Society, USA

[4.6] G.C. Smith and R. Blumenthal (*eds.*) (1991), "Tungsten and Other Advanced Metals for ULSI Applications VI", The Materials Research Society, USA

[4.7] V.V.S. Rana, R.V. Joshi and I. Ohdomari (*eds.*) (1992) "Advanced Metallization for ULSI Applications", The Materials Research Society, USA

[4.8] J.E.J. Schmitz, R.C. Ellwanger and A.J.M van Dijk (1988) "Characterization of process parameters for blanket tungsten contact fill" in "Tungsten and Other Refractory Metals for VLSI Applications III", V. Wells (*ed.*), pp. 55-61, The Materials Research Society, Pittsburgh, USA

[4.9] J.E.J. Schmitz, A.J.M. van Dijk, J.L.G. Suijker, M.J. Buiting and R.C. Ellwanger (1989) "A high throughput blanket tungsten process based on H_2/WF_6 chemistry" in "Proceedings of the European Workshop on Refractory Metals and Silicides", R. de Keersmaecker and K. Maex (*eds.*), pp. 350-358, North-Holland Publ. Comp., Amsterdam, The Netherlands

[4.10] R.V. Joshi, E. Mehter, M. Chow, M. Ishaq, S. Kang, P. Geraghty and J. McInerney (1990) "High growth rate CVD-W process for filling high aspect ratio sub-micron contacts/lines" in "Tungsten and Other Advanced Metals for VLSI/ULSI Applications V", S.S. Wong and S. Furukawa (*eds.*), pp. 157-166, The Materials Research Society, Pittsburgh, USA

[4.11] N.E. Miller and I. Beinglass (1980) "Hot-wall CVD tungsten for VLSI", *Solid State Techn.* **23**, pp. 79-82

[4.12] C.M. McConica and K. Krishnamani (1986) "The kinetics of LPCVD tungsten deposition in a single wafer reactor", *J. Electrochem. Soc.* **133** (12), pp. 2542-2548

[4.13] R. Blumenthal and G.C. Smith (1988) "Step coverage in coldwall -deposited blanket CVD tungsten films" in "Tungsten and Other Refractory Metals for VLSI Applications III", V. Wells (*ed.*), pp. 47-54, The Materials Research Society, Pittsburgh, USA

[4.14] E.K. Broadbent and C.L. Ramiller (1984), "Selective low pressure chemical vapor deposition of tungsten", *J. Electrochem. Soc.* **131** (6), pp.1427-1433

[4.15] P. van der Putte (1987) "The reaction kinetics of the H_2 reduction of WF_6 in the chemical vapor deposition of tungsten films" in "Tungsten and Other Refractory Metals for VLSI Applications II", E. Broadbent (*ed.*), pp. 77-84, The Materials Research Society, USA

[4.16] C.M. McConica and S. Churchill (1988) "Step coverage prediction during blanket CVD tungsten deposition" in "Tungsten and Other Refractory Metals for VLSI Applications III", V. Wells (*ed.*), pp. 257-262, The Materials Research Society, Pittsburgh, USA

[4.17] S. Chatterjee and C.M. McConica (1990) "Prediction of step coverage during blanket CVD Tungsten deposition in cylindrical pores", *J.Electrochem. Soc.* **137** (1), pp. 328-335

[4.18] J.E.J. Schmitz, W.L.N. van der Sluys and A.H. Montree (1990) "Comparison of calculated and experimental step coverage of the H_2/WF_6 and SiH4/WF_6 chemistries used in the blanket deposition of tungsten" in "Tungsten and Other Advanced Metals for VLSI/ULSI Applications V", S.S. Wong and S. Furukawa (*eds.*), pp. 117-124, The Materials Research Society, Pittsburgh, USA

[4.19] A. Hasper, C.R. Kleijn, J. Holleman and J. Middelhoek (1990) "W-LPCVD step coverage and modelling in trenches and contact holes" in "Tungsten and Other Advanced Metals for VLSI/ULSI Applications V", S.S. Wong and S. Furukawa (*eds.*), pp. 127-134, The Materials Research Society, Pittsburgh, USA

[4.20] A. Hasper, C.R. Kleijn, J. Holleman, J. Middelhoek and C.J. Hoogendoorn (1991) "Modeling and optimization of the step coverage of tungsten LPCVD in trenches and contact holes", *J. Electrochem. Soc.*, **138**, pp. 1728-1738

[4.21] T. Moriya, K. Yamada, Y. Tsunashima, S. Nakata and M. Kashiwagi (1983) "A new encroachment-free tungsten CVD process with superior selectivity" in "Extended Abstracts 15th Conf. Solid State Dev. and Mat.", pp. 225-228, Tokyo

[4.22] C.R. Kleijn, A. Hasper, J. Holleman, C.J. Hoogendoorn and J. Middelhoek (1990) "An experimental and modelling study of the tungsten LPCVD growth kinetics from H_2-WF_6 at low WF_6 partial pressures", in "Tungsten and other advanced metals for VLSI/ULSI applications V", S. Wong and S. Furukawa (*eds*), The Materials Research Society, Pittsburgh, USA, pp. 109-116

[4.23] C.R. Kleijn, A. Hasper, J. Holleman, C.J. Hoogendoorn and J. Middelhoek (1991) "Transport phenomena in tungsten LPCVD in a single-wafer reactor", *J. Electrochem. Soc.* **138**, pp. 509-517

[4.24] R. Arora and R. Pollard (1991) "A mathematical model for Chemical Vapor Deposition influenced by surface reaction kinetics: Application to low pressure deposition of tungsten", *J. Electrochem. Soc.* **138** (5), pp. 1523-1537

[4.25] W.A. Bryant (1978) "Kinetics of tungsten deposition by the reaction of WF_6 and hydrogen", *J. Electrochem. Soc.* **125** (9), pp. 1534-1543

[4.26] E.K. Broadbent and W.T. Stacy (1985) "Selective tungsten processing by low pressure CVD", *Solid State Techn.* **28**, pp. 51-59

[4.27] Y. Pauleau and Ph. Lami (1985) "Kinetics and mechanism of selective tungsten deposition by LPCVD", *J. Electrochem. Soc.* **132** (11), pp. 2779-2784

[4.28] J.I. Ulacia F., S. Howell, H. Körner and Ch. Werner, (1989), "Flow and reaction simulation of a tungsten CVD reactor", *Appl. Surf. Science*, **31**, pp. 370-385

[4.29] M. Knudsen (1950), "The kinetic theory of gases", J. Wiley & Sons, New York, USA

[4.30] S. Dushman (1962), in: J.M. Lafferty (*ed.*) "Scientific foundations of vacuum technique", J. Wiley & Sons, New York, USA

Chapter 5

Selective tungsten deposition

5.1   Introduction

In this chapter simulation results of selective tungsten deposition in a cold wall single wafer reactor will be presented. This process has received widespread attention in past years [5.1, 5.2, 5.3, 5.4, 5.5, 5.6], since it potentially offers drastic process simplification for filling vias and contact holes with a highly conductive tungsten plug without depositing the metal on adjacent silicon oxide regions. Nevertheless today there are still a large number of processing problems which have prevented so far the general acceptance of selective tungsten in IC manufacturing despite its unquestioned advantages concerning cost and process simplifications.

The reason for this lack of acceptance is the questionable reliability in maintaining good selectivity without the formation of tungsten islands on the oxide regions which ideally should be non reactive. A number of possible influences on selectivity loss have been investigated in the literature. On the one hand it is well accepted that careful cleaning and preprocessing steps can minimize the number of possible nucleation sites on the oxide surfaces [5.7]. On the other hand it has been shown that there are highly reactive intermediates, which are produced in low concentrations during the tungsten deposition process on the reactive metallic surface and can be transported through the gas phase to the non reactive surface parts [5.2]. Here they can give rise to nucleation of tungsten islands and initiate the loss of selectivity.

Of course the type of macroscopic simulation presented in this book cannot address the influence of surface cleanliness on selectivity, but it should be able to contribute to the understanding of the generation and transport of those reactive intermediates and the influence of their concentration on selectivity loss.

The purpose of this chapter is to present a numerical design tool which could help equipment designers and process engineers to better understand the reasons for selectivity loss and establish a reliable process for selective tungsten deposition in a manufacturing environment. To this end the models from chapter 4 will be enhanced to include the reactive intermediates WF_x and SiF_y, and a model for the autocatalytic growth of tungsten nuclei will be presented. Model results are in good agreement with experimental findings. Simulations are performed to show the basic consequences of these models with two different chemical models for selectivity loss featuring the role of tungsten- and silicon-subfluorides, respectively. Moreover the simulations are used to study the impact of reactor design on selectivity.

5.2   Chemical Reactions

5.2.1   Deposition on Tungsten Surfaces

In principle the simulation of selective tungsten deposition can proceed very similarly to the blanket deposition as described in chapter 4, when the fraction of reactive surface area (contact holes, metallic surfaces etc.) is taken into account.

On a surface covered with tungsten the deposition occurs by a reduction of tungsten hexafluoride through the overall reaction (4.1)

$$WF_6(g) \; + \; 3\,H_2(g) \; \longrightarrow \; W(s) \; + \; 6\,HF(g) \,.$$

For the reaction rate we use equations (4.2) and (4.3) together with the boundary condition (2.45). Because the reaction only takes place on the tungsten covered fraction of the surface and not on the regions covered with SiO_2, the reaction rate from (4.3) was multiplied with the factor

$$\theta \; = \; \frac{Area\ covered\ with\ tungsten}{total\ wafer\ area} \tag{5.1}$$

The area fraction θ is a function of position across the wafer, but will always be represented by an average value over several mean free paths

($\approx 0.1mm$) and thus can never resolve individual structures in a μm scale, such as contact holes. However,a macroscopic loss of selectivity giving enhanced density of nuclei (e.g. at the wafer edges), can be adequately described by these simulations.

5.2.2 Deposition on Silicon Oxide Surfaces

Two main models for selectivity loss are found in recent publications, featuring the role of tungsten subfluorides WF_x and silicon subfluorides SiF_y, respectively.

WF_x Model

In [5.2] Creighton investigated tungsten deposition in an experimental ultra high vacuum chamber using Auger electron spectroscopy and temperature programmed desorption experiments. He suggests the importance of tungsten subfluorides WF_x with $x \leq 5$ which are formed as intermediates in the WF_6 reduction on the metallic surface and are transported through the gas phase to SiO_2 covered regions. Due to their higher reactivity, WF_x molecules will induce tungsten nucleation and produce the observed selectivity loss.

The kinetic process was studied theoretically by Arora and Pollard [5.9] and experimentally by Creighton [5.2]. A brief description of the mechanism is seen in Fig. 5.1.

In the following description, the exponent in parenthesis for a variable $A_i^{()}$ represents the phase adsorbed on tungsten (W), adsorbed on silicon dioxide (SiO_2), or in the gas phase (g). Here, $WF_6^{(g)}$ is transported to the surface by convection and diffusion identified by the double arrow labeled (1). The concentration of $WF_6^{(g)}$ in the gas is computed from a fluid flow solution at the operating conditions as described previously, and at the surface $WF_6^{(W)}$ is proportional to $WF_6^{(g)}$ and the relative tungsten area θ_W. Processes (2) and (3) represent the decomposition of WF_6 with H_2 to create tungsten as described in chapter 4 and are illustrated with double arrows to identify the most important reaction paths. In the fragmentation process, the model assumes that one intermediate species is able to desorb or sublimate [5.2, 5.3]. The only possible candidates must be highly fluorinated tungsten compounds (WF_5 or WF_4) that have high vapor pressures. The precise nature of the tungsten species is not important for the model because this compound is generated in small quantities

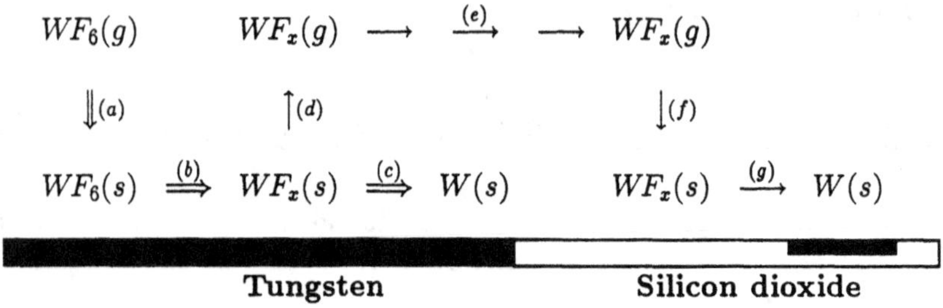

$$WF_6(g) \qquad WF_x(g) \longrightarrow \xrightarrow{(e)} \longrightarrow WF_x(g)$$

$$\Big\Downarrow(a) \qquad\qquad \Big\uparrow(d) \qquad\qquad\qquad \Big\downarrow(f)$$

$$WF_6(s) \xrightarrow{(b)} WF_x(s) \xrightarrow{(c)} W(s) \qquad\qquad WF_x(s) \xrightarrow{(g)} W(s)$$

Tungsten   **Silicon dioxide**

Figure 5.1: Reaction mechanism for the loss of selectivity. The double arrows show the main reaction path and the letters represent (1) adsorption of WF_6 on the surface, (2) the chemical reaction to form the tungsten fluorinated compound WF_x, (3) fragmentation reaction to form tungsten, (4) desorption of WF_x into the gas, (5) diffusion and convection in the gas phase, (6) adsorption of WF_x on Silicon dioxide, and (7) chemical reaction to form W on SiO_2.

and does not introduce large errors in the computation of the mass- and particle-continuity equations.

In their theoretical study Arora and Pollard [5.9] derived results for the partial pressure of WF_5 and WF_4 above the reacting surface. Replotted on a linear scale their results are given in Fig. 5.2 and Fig. 5.3. It is outside the scope of our investigation to model the interdependencies between the WF_5 and WF_4 components on the surface and in the gas phase, but from Fig. 5.2 we can deduce that both molecules are generated primarily at the surface and not in the gas phase. Moreover Fig. 5.3 shows that the sum of partial pressures for WF_5 and WF_4 above the surface is approximately proportional to the WF_6 partial pressure.

To avoid the kinetic description for the reduction of WF_6 and because the reaction is kinetically limited by some intermediate step, it will be assumed that the desorption rate of the fragment on the tungsten is proportional to $WF_6^{(g)}$ at the boundary. In the rest of this chapter we will use the notation WF_x for the reactive fragments and treat them as a single species. In the gas, the intermediate is transported like any other gaseous species (5), and its concentration is computed from a chemical-species-continuity equation with a boundary flux that includes desorption and deposition on both

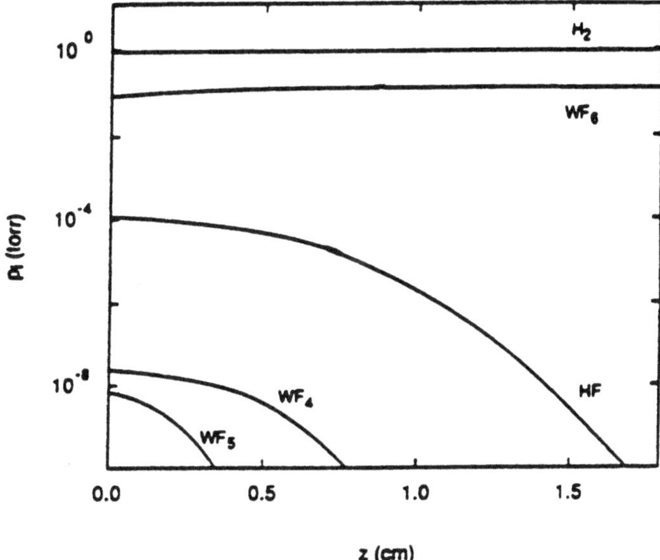

Figure 5.2: Partial pressures of different chemical species as a function of distance above the substrate. (after [5.9])

tungsten and silicon dioxide.

If the fragment deposits on tungsten, the reaction proceeds normally and creates more selective tungsten; however, if it deposits on silicon dioxide (6), it disproportionates creating tungsten nucleates (7). Therefore we can model this process as

$$j_{WF_x} = K_a\,\omega_{WF_6}\,\theta - K_b\,\omega_{WF_x} \tag{5.2}$$

$$n = K_c\,\omega_{WF_x}(1-\theta) \tag{5.3}$$

j_{WF_x} is the net surface flux of WF_x, caused by adsorption of WF_x on the whole surface (adsorption coefficient K_b) and by desorption of the WF_x intermediates in the WF_6 reduction chain, which occurs only on tungsten covered parts of the surface (fraction θ). In equation (5.3) n is the nucleation rate of tungsten nuclei on the SiO_2 surface (fraction $1-\theta$).

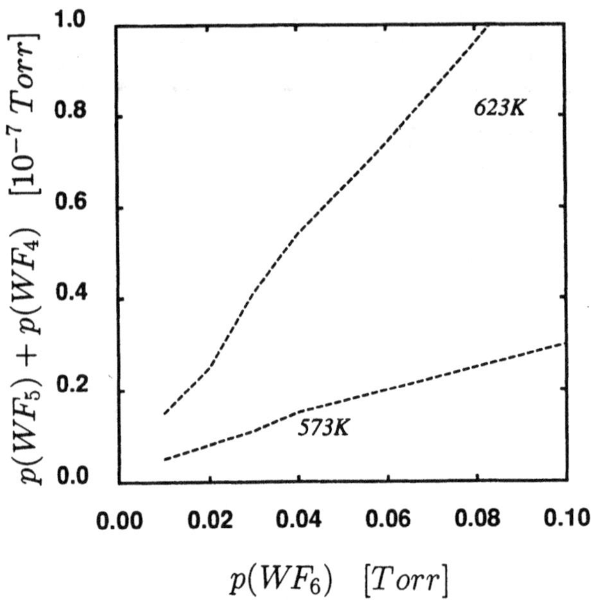

Figure 5.3: partial pressure of WF_x as a function of WF_6 concentration (replotted from data in [5.9]).

With the three parameters K_a, K_b, K_c it is possible to get a qualitative agreement with published nucleation data [5.1, 5.3, 5.6]. It is clear, however, that all these parameters would be strongly influenced by temperature, and at least the nucleation rate K_c also by surface cleanliness and wafer history [5.7]. In this work we have used $K_a = 10^{-5}\,kgs^{-1}m^{-2}$, $K_b = 10^{-2}\,kgs^{-1}m^{-2}$, and $K_c = 10^{13}s^{-1}m^{-2}$ which result in reasonable nucleation behaviour for the examples studied.

An improved approach should use the concentrations c_{WF_6} instead of the mass fractions ω_{WF_6} etc. in Equ. 5.2 and 5.3, but since the mass fractions are the variables solved in the differential equations, we have chosen the ω_i's in our model. This could make the coefficients $K_a - K_c$ concentration dependent, if the WF_6 mass fraction is varied over a wide range, but for the small variations studied in this work we can keep them as constant.

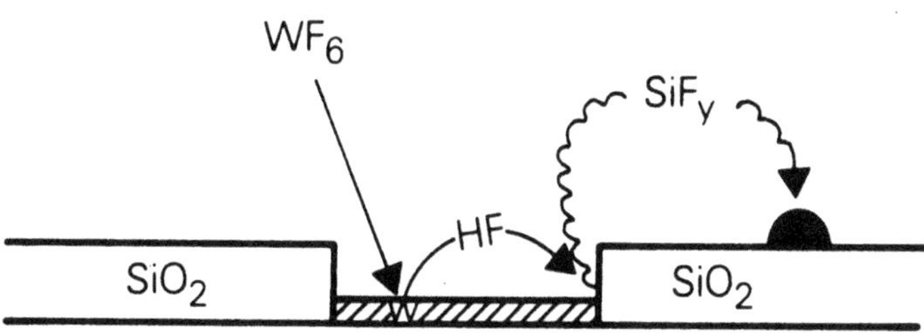

Figure 5.4: Alternative chemical model for selectivity loss, featuring the role of silicon subfluoride.

SiF_y Model

A different mechanism was proposed by McConica [5.1] and then strongly supported by the experiments of Hirase et al. [5.8]. It is hypothesized that HF reacts with SiO_2 to form SiF_y (y$\leq$ 3) in the presence of a tungsten catalyst, as shown in Fig.5.4. The highly reactive species SiF_y then travels through the gas phase, reacts with WF_6, and leaves WF_x, which forms a surface nucleus. Once WF_x is adsorbed, it forms a continuously growing tungsten island.

The model proposed in [5.1] can be formulated similarly to (5.2) and (5.3).

$$j_{SiF_y} = K_d \, (\omega_{HF})^\nu \, \theta - K_e \, \omega_{SiF_y} \tag{5.4}$$

$$n = K_f \, \omega_{SiF_y}(1 - \theta) \tag{5.5}$$

There are two basic differences between the models. The SiF_y model suggests a two step process. First HF is produced on all hot tungsten surfaces, while in the second step the HF reacts along the boundaries of tungsten islands with SiO_2 or during the initial step of tungsten growth with the underlying silicon atoms to form the actual selectivity killer species SiF_y. Moreover this model proposes a higher reaction order ν in the production of the intermediate. Kwakman [5.5] has measured the influence of HF concentration on nucleation rates and found a very high nonlinearity of $\nu = 6 - 12$. In our calculations we used

$K_d = 10^{29} kgs^{-1}m^{-2}, K_e = 10^{-2} kgs^{-1}m^{-2}, K_f = 0.5 \cdot 10^{14} s^{-1}m^{-2}$ and $\nu = 12$ to get nucleation rates comparable to equation (5.3).

We assume that the contact holes and all the growing nuclei on the wafer are small enough compared with a surface diffusion length of HF so that all molecules produced in the contact holes can reach a SiO_2 boundary and produce SiF_y according to Fig. 5.4. Thus we can make the SiF_y production proportional to the tungsten area ratio θ. This however is not true when $\theta \approx 1$ as on the susceptor, where we have set the SiF_y production to zero.

Many observations are reported in literature which could favour one model or the other. We will see in section 5.3 that the two models result in very different concentration profiles for SiF_y and WF_x, respectively, however, the actual nucleation rates behave rather similarly and can be fitted to experimental data for both models, if the proper parameters are chosen in equations (5.2)-(5.5).

5.2.3   Selectivity-loss Model

In this section we want to establish a model for the time dependence of the tungsten covered area fraction θ to allow a consistent mathematical formulation of selectivity-loss, which can be used in a detailed 2-D fluid flow simulation.

If a nucleus of radius r grows laterally at rate $\mathcal{G}$, its area increases by dA during time dt such that,

$$dA = 2r\pi \cdot \mathcal{G}dt$$

The actual radius of a nucleus at time t is denoted by $r(t, t_0)$, when the nucleus was generated at time t_0 and is described by

$$r(t, t_0) = \int_{t_0}^{t} \mathcal{G}(t') \, dt' \tag{5.6}$$

where the initial radius of the nuclei is neglected as being small compared to the final radius.

The total area growth is found from integrating over all nuclei with different size

$$\frac{dA}{dt} = \int_{0}^{t} 2r(t, t_0)\pi \cdot \mathcal{G} \cdot n(t_0) \cdot A_{tot} \cdot dt_0 \tag{5.7}$$

where $n(t_0)$ is the nucleation rate per unit area at the time t_0, when the nucleus was generated, as given by Eq.(5.3) or (5.5).

Equation (5.7) applies only if nuclei do not overlap, i.e. when $\theta \ll 1$. As nuclei begin to overlap, the area growth is decreased such that A asymptotically approaches the total wafer area A_{tot}. As this region is not critical for selectivity loss investigations, an empirical factor $(1 - \theta)$ was added to (5.7), which gives the correct asymptotic behaviour for both $A \ll A_{tot}$ and $A \approx A_{tot}$. We do not feel it is necessary to implement a more sophisticated percolation model for this study.

Inserting (5.6) into (5.7) and dividing by A_{tot}

$$\frac{d\theta}{dt} = (1 - \theta)2\pi\mathcal{G}(t) \int_0^t n(t_o) \int_{t_o}^t \mathcal{G}(t')dt'dt_o \tag{5.8}$$

utilising $\theta = A/A_{tot}$

Reversing the order of integration we get,

$$\frac{d\theta}{dt} = (1 - \theta)2\pi\mathcal{G}(t) \int_0^t \mathcal{G}(t') \cdot N_{tot}(t')dt' \tag{5.9}$$

with the total number of deposited nuclei

$$N_{tot}(t) = \int_0^t n(t')dt'$$

In the absence of any more appropriate assumptions the vertical deposition rate of equations (4.2) and (4.3) and (2.33)was also used for the lateral nuclei growth rate $\mathcal{G}$.

Assuming constant deposition $\mathcal{G}$ and nucleation rate n, we can simplify the equation to

$$\frac{d\theta}{dt} = (1 - \theta)\pi\mathcal{G}^2nt^2$$

for which an analytical solution can be given.

$$\theta(t) = 1 - (1 - \theta(0))\exp(-at^3) \tag{5.10}$$

with $a = \frac{1}{3}\mathcal{G}^2\pi n$. Fig. 5.5 shows the behaviour of Equ. 5.10 for different values of the parameter a. This function behaves as t^3 for small t's and approaches $\theta = 1$ as t gets large. However, we will see in the next section that n is not constant in time so that this simplified solution does not describe the accurate solution.

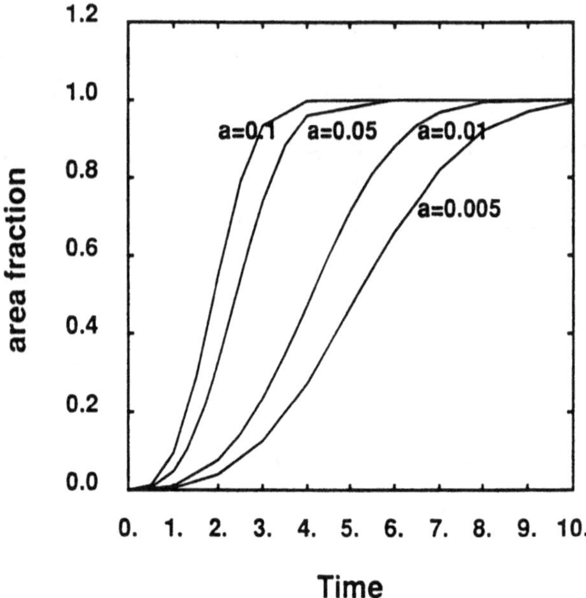

Figure 5.5: Analytical solution for the area fraction θ of the wafer that is covered with tungsten. The island growth rate and the nucleation rate are assumed time independent.

5.3   Simulation Results

5.3.1   Numerical Aspects

The results to be described in this section were derived using the commercial fluid flow simulator PHOENICS [5.10] which solves the transport equations (2.1)-(2.7) from chapter 2 using the hybrid scheme discretization (2.105) together with the SIMPLE algorithm described in section 2.9.2 and the line-by-line Gauss-Seidel method from section 2.9.5.

Appropriate subroutines have been added PHOENICS to implement the transport properties of the five component gas mixture according to the formulae given in section 2.7.

For the sake of simplicity the thermal diffusion coefficients were not implemented in the full form of Equ. (2.89) but we rather used an approximation taking the thermal diffusion factor α_{ij} in Equ. (2.88) as a constant parameter. The values for α_{ij} were estimated from Equ.(2.87). Comparison with more sophisticated models [5.12] have shown that these numbers are realistic estimates as long as $\omega_i \leq 0.5$. We feel that our approach gives an upper bound for the effect of thermodiffusion, since the rigid elastic sphere model used in Equ. (2.87) is known to overestimate thermal diffusion.

In the following simulations the three species HF, WF_x, and SiF_y are present only in low mass and mole concentrations. For this case the multicomponent currents of Equ.(2.8), (2.15) and (2.19) can be written as

$$\underline{j}_2 = -\rho D_{21}(\nabla\omega_2 + \omega_2\omega_1\frac{\alpha_2}{T}\nabla T) \ \ for \ 1 = "H_2", \ 2 = "WF_6" \quad (5.11)$$

$$\underline{j}_i = -\rho\mathbb{D}_i(\nabla\omega_i + \omega_i[\beta_i\nabla\omega_2 + \frac{\alpha_i}{T}\nabla T]); \ \ i = 3,4,5; \ \omega_i \ll 1 \quad (5.12)$$

The form of the cross diffusion ratio β_i can be found by a straightforward derivation from equation (2.12), using $\omega_i \ll 1$; $i \geq 3$, and gives:

$$\beta_i = \frac{m}{m_1}(1 - \frac{D_{12}}{D_{1i}}) - \frac{m}{m_2}(1 - \frac{D_{12}}{D_{2i}}) \quad (5.13)$$

The effective diffusion coefficient $\mathbb{D}_i$ of species i into the mixture is given by Equ. (2.16).

At each time step the equation (5.9) was solved in a fully implicit discretized formulation using the nucleation rate models (5.3) or (5.5) respectively. The boundary conditions of the diffusion equations (5.2) and (5.4) were evaluated using θ from (5.9). To consider the mutual dependencies of boundary conditions and transport parameters of all the equations, the whole system was iterated at each time step until a self-consistent solution was obtained. All the equations have been solved fully time dependent as described in Eq. (2.1)-(2.7) and (5.9).

The time steps in the numerical solution have been chosen small enough that the solution will not depend on the time step length. This requires time steps as small as $10^{-3}s$ at the beginning when WF_6 is switched on. As time proceeds the steps can be gradually enlarged up to $300s$.

The model was applied to simulate selective tungsten deposition in a cold wall single wafer reactor, shown in Fig. 5.6. In the reactor simulated,

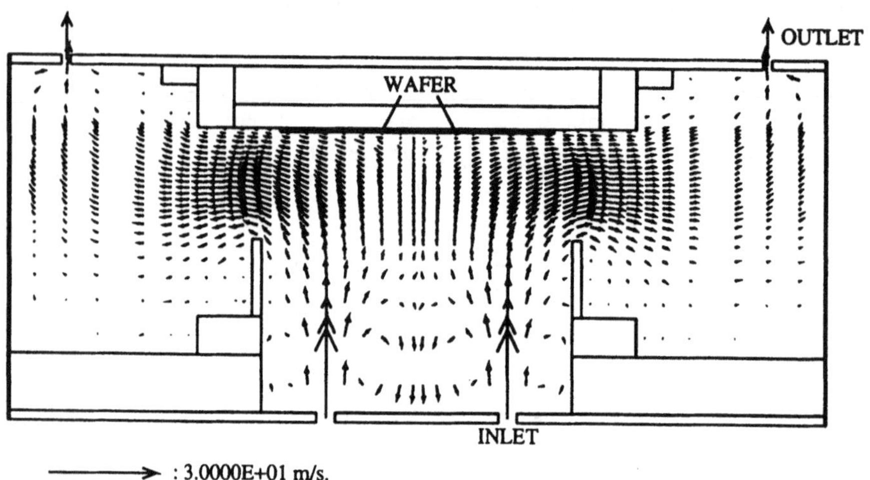

Figure 5.6: gas flow velocity in the simulated CVD reactor. Except in the inlet region there is a perfectly homogenous flow pattern.

the wafer faces down onto the gas flow and is heated from the rear with halogen lamps. The gas is introduced at the bottom of the reactor and extracted at the top through four pumping ports. Due to the symmetry inherent to the problem it was possible to restrict the simulation to two-dimensional geometry on a cylindrical grid. The spatial discretization uses a grid spacing of 2 mm near the wafer, which is gradually enlarged up to 3 cm towards the outer reactor walls. Care was taken not to influence the distribution near the wafer by coarse discretization, but we did not make much effort to resolve the flow pattern near inlet and outlet. The grid used in this investigation is shown in Fig. 5.7. Typical wafer temperature was $550 - 700K$ and pressure $400 - 900mTorr$. Gas inflows were taken to be $1 - 2slm$ for hydrogen and $10 - 100sccm$ for WF_6.

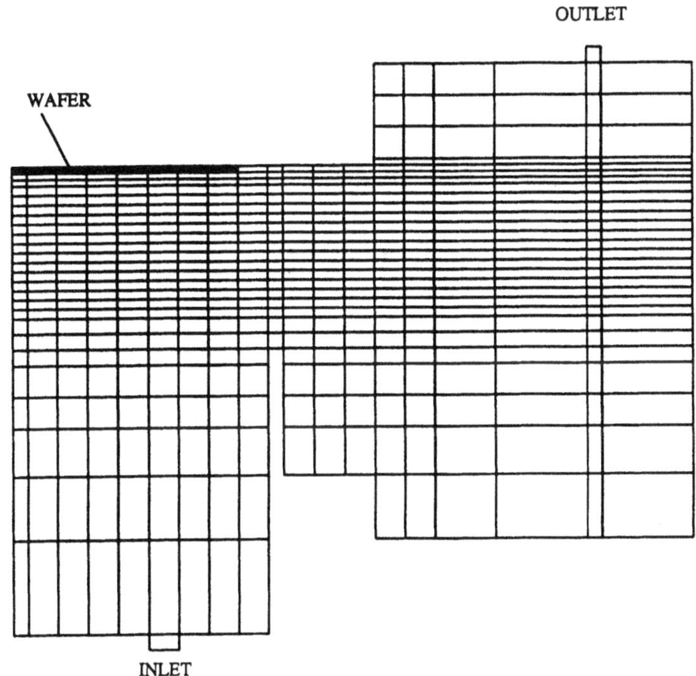

Figure 5.7: discretization used in the simulation of the selective tungsten CVD reactor.

5.3.2 Time dependence

As a starting point, a wafer with a reactive area fraction of $\theta = 0.01$ was assumed. The wafer was surrounded by a hot susceptor plate, fully covered with tungsten from earlier depositions ($\theta = 1$). The flow calculations start with a stationary situation, in which the reactor is flooded with 100 % hydrogen. At the start of processing time the WF_6 flow is abruptly switched on at the inlet and it turned out that the reaction gas fills the chamber within 1 second. In our model all the reaction products approach their quasistationary distributions within the first 5-10 seconds. During this time the number of deposited nuclei is very small, so that the area fraction θ stays very closely at its initial value. Only after several minutes θ begins slowly to grow. This in turn changes the boundary conditions for the diffusion equations and induces a time dependent variation of the concentration profiles in the reactor.

After this behaviour had been verified for some cases, the calculations were started from the quasistationary distributions with WF_6 gas flowing and the area fraction θ on its initial value. In those calculations time steps of 0.5 to 5 minutes could be used.

5.3.3 Concentration Profiles

The gas flow velocity in the reactor can be seen in Fig. 5.6. Some recirculation flow pattern can be identified at the inlet but a perfect homogeneous flow is established near the wafer. Fig. 5.8-5.13 give temperature and concentration contours in the right half of the reactor from Fig.5.6. They were plotted for a steady state condition, where the deposition rate and all the concentrations had reached a quasistationary distribution, but the number of nuclei is still small enough, so that $\theta \approx \theta(0)$.

The temperature contours giving the temperature gradient from the hot susceptor to the cold walls at inlet and outlet are found in Fig. 5.8. Figure 5.9 illustrates the mole fraction contribution for the reactive gas WF_6. It enters the reactor with the specified concentration of 5% and is decreased due to the deposition at the wafer. Because of the 100% reactive surface at the susceptor more WF_6 is consumed outside the wafer area than above the wafer. Clearly observable is the effect of thermodiffusion, which leads to enhanced concentration values in the cold regions of the reactor. Moreover a pronounced concentration maximum builds up against the flow direction below the wafer, where the convective inflow from the inlet to the wafer and the thermodiffusion flux in the opposite direction compensate.

In Fig.5.10 the concentration profile of HF is shown. This species is produced as a reaction product, mainly at the hot susceptor outside the wafer area and to a smaller extent on the wafer surface. Because of its high diffusion coefficient HF is distributed virtually evenly over the whole reactor, only at the inlet there is a pronounced minimum resulting from the convective inflow of the HF free inlet gas. The maximum at the outer wall comes from the thermal diffusion current to the cold reactor walls. Note the gradient of HF concentration along the surface of the wafer, which leads to a factor 2 reduction between the periphery and center. This is due to the higher production of HF at the fully tungsten covered susceptor in comparison to the 1% covered wafer.

The concentration profile of the carrier gas H_2, shown in Fig.5.11, is basically the opposite of the WF_6 profile. Mole fractions are mostly above 0.90 with a maximum at the hot wafer and minima at the cold walls.

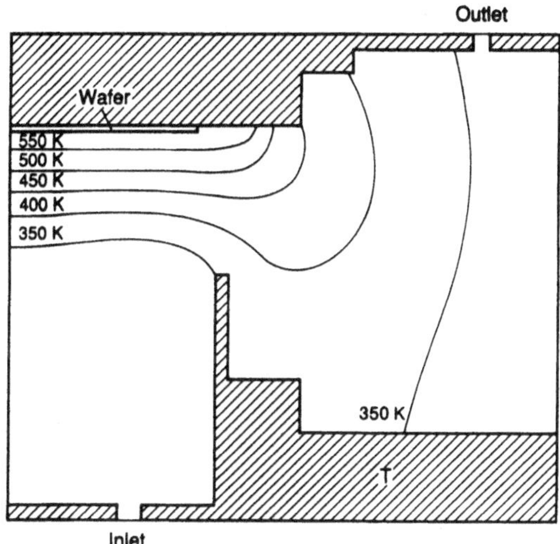

Figure 5.8: Temperature contours inside the CVD reactor. The temperature increases as the gas approaches the hot wafer.

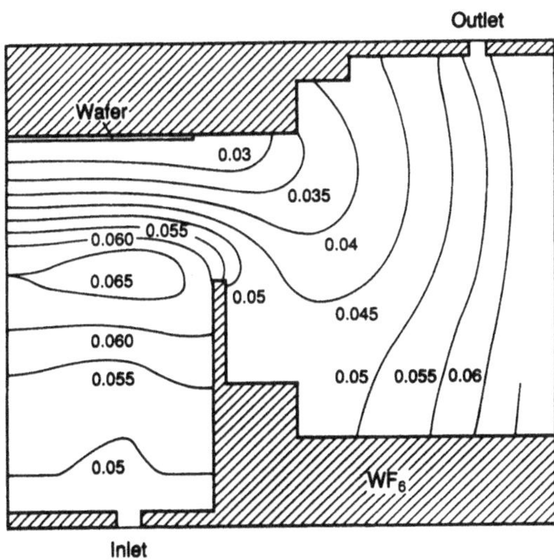

Figure 5.9: Concentration contours of tungsten hexafluoride inside the reactor. Thermodiffusion leads to pronounced maxima at the inlet and near the cold walls.

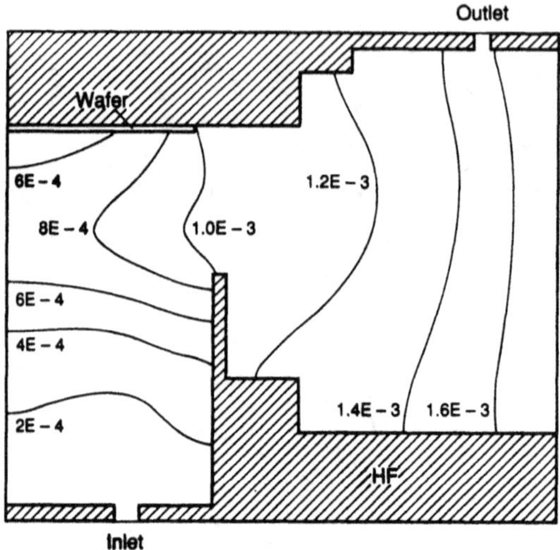

Figure 5.10: Concentration contours of hydrogen fluoride inside the reactor. *HF* is produced at the hot susceptor around the wafer and diffuses through the whole reactor.

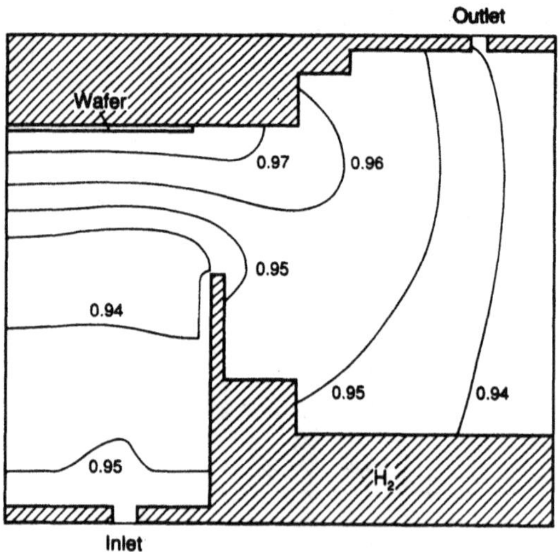

Figure 5.11: Concentration contours of hydrogen inside the reactor. H_2 accumulates at the hot wafer due to the thermodiffusion effect.

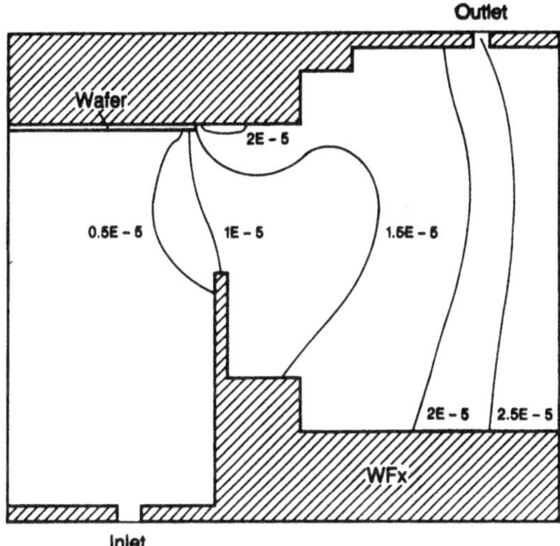

Figure 5.12: Concentration contours of WF_x inside the reactor. WF_x desorbs from the hot susceptor as a byproduct of the WF_6 reduction and is adsorbed at the wafer surface.

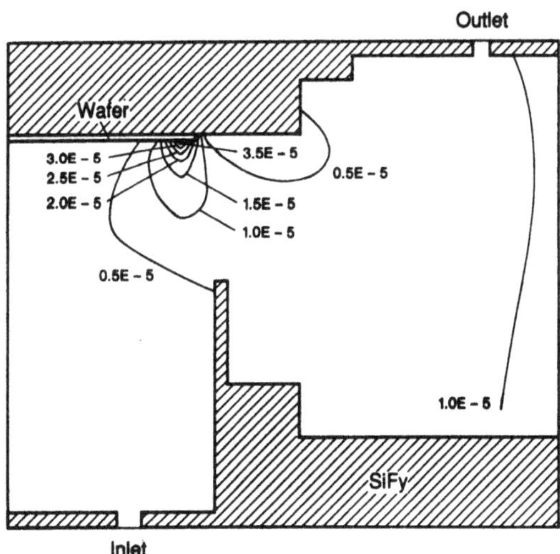

Figure 5.13: Concentration contours of SiF_y inside the reactor. This species is produced as a secondary reaction product in the wafer periphery, where the HF concentration is high.

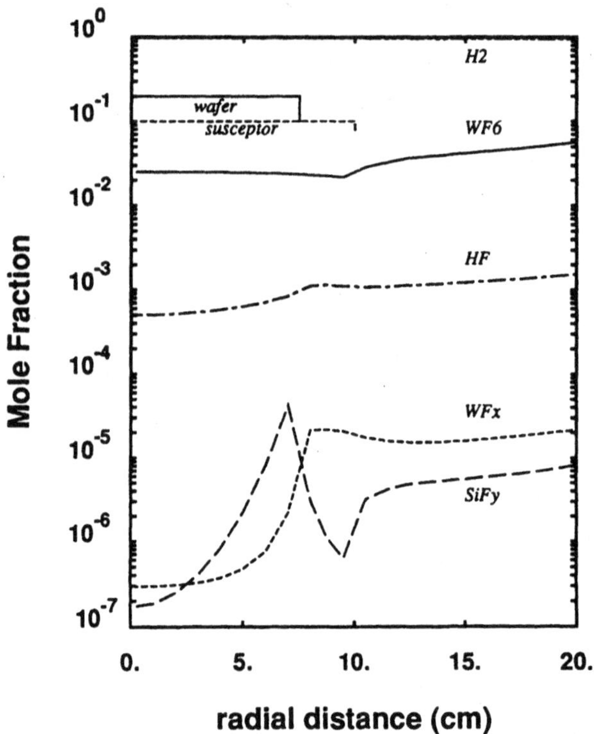

Figure 5.14: Concentration profiles of all chemical species along a radius through the reactor immediately above the wafer surface.

Fig. 5.12 gives the concentration of the byproduct WF_x as calculated by equ.(5.2). It has some similarities with HF, only the absolute values of concentration are much lower. We find a similar concentration maximum at the cold outlet and a minimum at the inlet. There is, however, a much larger gradient across the wafer surface, which gives a drop in excess of factor 10 between edge and center. Unlike HF, WF_x is consumed at the SiO_2 surface, leading to tungsten nucleation and selectivity loss. The small amount of WF_x produced at the susceptor ring is readily consumed at the SiO_2 surfaces leading to the observed concentration drop.

Concentration contours for SiF_y as determined from equation (5.4) are shown in Fig. 5.13. For its production HF, SiO_2 and a catalytic tungsten surface are required. So there is no SiF_y production at the susceptor (pure tungsten surface), only at the wafer, where both SiO_2 and tungsten surfaces are present close together. The production rate follows the HF concentration gradient, leading to a maximum of SiF_y at the wafer edge decreasing towards the wafer center and also along the susceptor. Because of the high nonlinearity in (5.4) the SiF_y gradient is much larger than the corresponding HF gradient.

Fig.5.14 gives concentration profiles across the wafer surface for all the species to allow the investigation of their mutual dependencies. All the concentrations increase at the cold outlet side, WF_6 is used up at the susceptor, where WF_x and HF are produced. Both SiF_y and WF_x are consumed at the wafer, with their gradients being approximately equal because the parameters of both models have been adjusted to fit the same set of experimental data (see Fig.5.16).

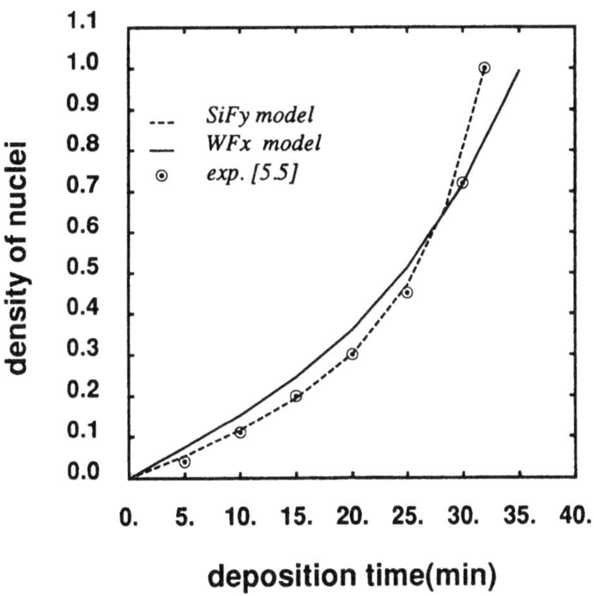

Figure 5.15: Density of tungsten nuclei as a function of deposition time. The autocatalytic role of the tungsten surface leads to an exponential increase. Both models can reproduce the experimental data from [5.5]

5.3.4   Nucleation Data

Using Eq.(5.2) and (5.3), nucleation rates were derived from these results for each point on the wafer and at all instants in time. To calibrate our model the parameter K_c in equ.(5.2) was adjusted to fit the experimental results given in [5.5]. However, it is well known that the nucleation depends strongly on surface impurities and pretreatment and thus we expect a strong dependence of K_c on the temperature and on the quality of the wafer. Nevertheless, the time dependence of the normalized curve in Fig.5.15 shows very good agreement between theory and experiment. A similar adjustment can also be made for the SiF_y based model of equ.(5.4) and (5.5). This is not surprising, since the time dependence of nucleation is governed by equ. (5.9) for both models.

The value K_b of the adsorption rate has a strong influence on the slope of the nucleation rate near the wafer edge. Fig. 5.16 shows the density of nu-

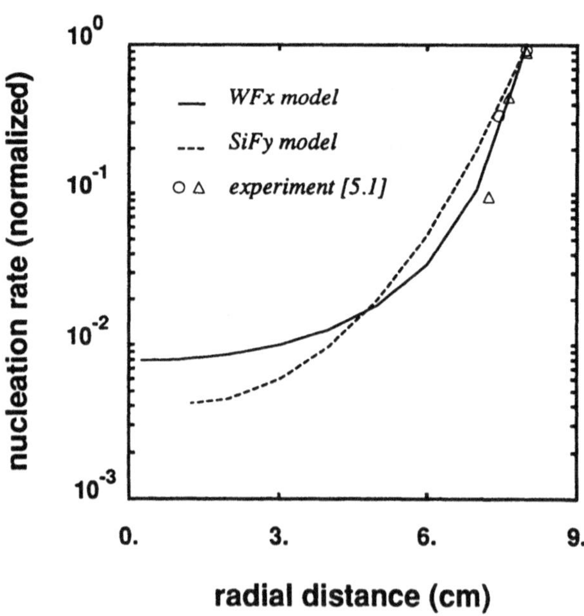

Figure 5.16: Nucleation rate along a wafer radius. Both models can be adjusted to the experimental results from [5.1].

clei along a wafer radius compared with experimental results of McConica [5.1, 5.3]. We see that we get reasonable fits to the data points for both models, however, more than 2 cm away from the wafer edge, where no data are available, the SiF_y model predicts significantly lower nucleation rates than the WF_x model.

In Fig. 5.17 and 5.18 the radial distributions of nuclei for the two models is compared. For both cases a pronounced increase towards the wafer edge is apparent, the selectivity killer species (WF_x and HF resp.) being produced at the hot susceptor plate. However, the SiF_y model predicts much steeper profiles, which are a result of the lower nucleation rate in the wafer center as shown in Fig. 5.16.

Qualitatively, the enhanced nucleation at the wafer edge has also been found experimentally [5.1, 5.3, 5.4, 5.6], but the data published so far do not allow to distinguish clearly between the models.

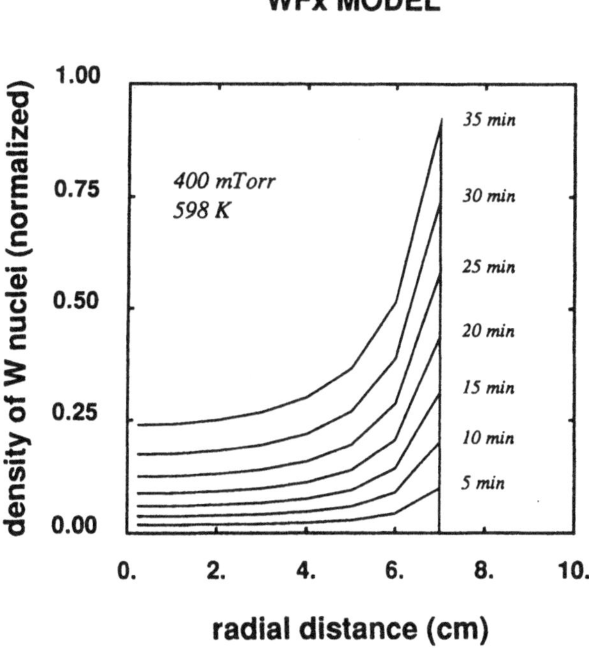

Figure 5.17: Density of tungsten nuclei on SiO_2 (normalized) as a function of position and time for the WF_x model.

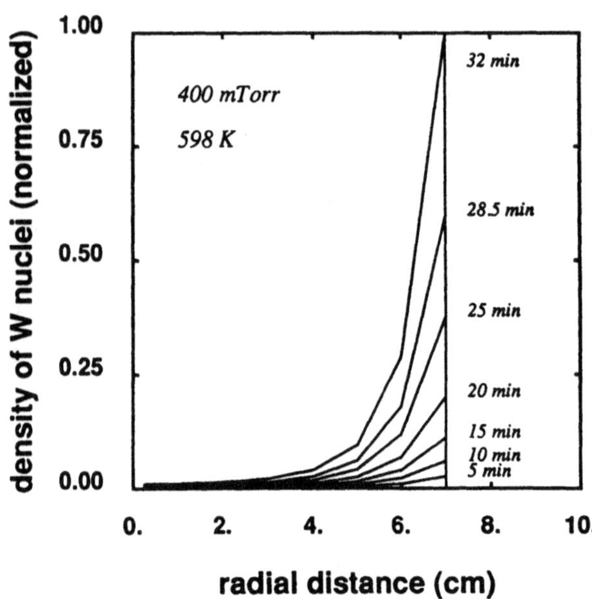

Figure 5.18: Density of tungsten nuclei on SiO_2 (normalized) as a function of position and time for the SiF_y model.

Pattee et al. [5.4] measured the velocity of the moving tungsten front from the wafer periphery towards the center on a polyimide covered wafer and found values between 0.1 and 0.3 cm per minute. This compares favorably with the ≈ 0.12 cm/min velocity of the steep front moving from the wafer edge towards the center in Fig. 5.17 and 5.18. However, in Pattee's investigation the transition to a completely covered tungsten surface seems to be much more abrupt than in our model. This is not surprising, since our model tries to describe the nucleation behaviour at low nuclei counts, while it is very crude in the percolation region.

Fig. 5.19 shows the time development of the area fraction θ covered with tungsten. The WF_x model has been taken in this case, while for the SiF_y model we would expect steeper gradients and a larger lag between edge and center curves. Unfortunately the calculations with the SiF_y model could not be continued for θ values above 0.3 due to convergency problems with the highly nonlinear boundary condition (5.4).

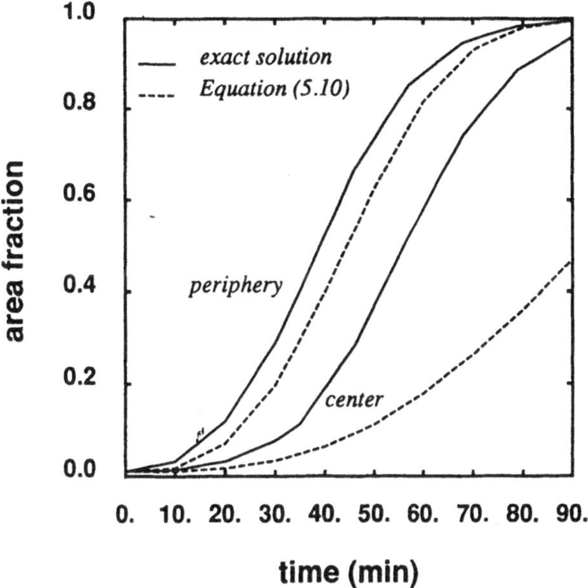

Figure 5.19: Area fraction of wafer surface, which is covered by tungsten. The autocatalytic role of tungsten leads to an exponential growth at the beginning, while after longer times the value 1 is approached. On the periphery, which lies adjacent to the hot susceptor, the increase is more rapid than in the center.

Several significant features can be observed in the curves of Fig. 5.19. In the initial phase we see an exponential growth, which is a consequence of the autocatalytic nucleation behaviour, because the growing nuclei themselves produce the selectivity killer species WF_x.

The time constant of the initial growth is smaller in the wafer center as compared to the edge. This behaviour is a consequence of the gradient of WF_x across the wafer surface, which can be seen from Fig. 5.14. At larger times we find a saturation at $\theta = 1$, which comes from overlapping of the growing nuclei, when their density is high.

Also compared was the analytical solution (5.10) using values of n and $\mathcal{G}$ from the steady state solution. Since this approach does not self consistently consider the enhanced WF_x production from the growing tungsten surfaces, it shows a reduced area growth, but the qualitative behaviour as a function of time comes out correct.

5.3.5   Design Variations

Fig.5.20 illustrates the influence of a modified reactor design, where the outlet gap at the wafer edge was reduced from 60 mm to 6 mm. This causes an increased gas speed near the hot susceptor which drives the byproduct HF away from the wafer. Therefore its concentration near the wafer is reduced by a factor 2-3 compared with the original design, as it is shown in Fig. 5.21. Outside the wafer at the susceptor the concentration is approximately equal for both designs.

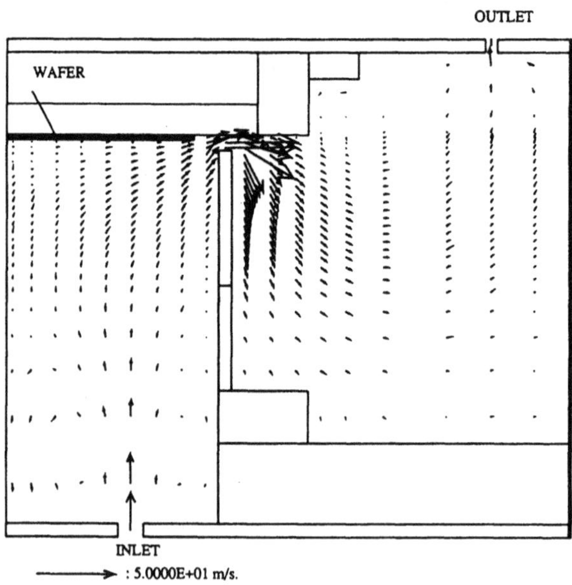

Figure 5.20: Gas velocity distribution for a reduced outlet gap of 6 mm. The small gap increases the gas velocity near the susceptor compared to figure 5.6.

Because of the high reaction order $\nu = 12$ in (5.4) the moderate reduction in HF is magnified several orders of magnitude if we consider the species SiF_y. On the other hand, in contrast to HF, the WF_x profile above the wafer is hardly influenced by the modification in reactor geometry. This has its reason in the very high reactivity of WF_x, which leads to a complete consumption of the molecules before they can be transported through the gas phase. Therefore a change in the convective transport velocity does not influence the concentration.

Since the nucleation rate is directly proportional to the concentrations of WF_x and SiF_y, respectively, we conclude that the two models will predict

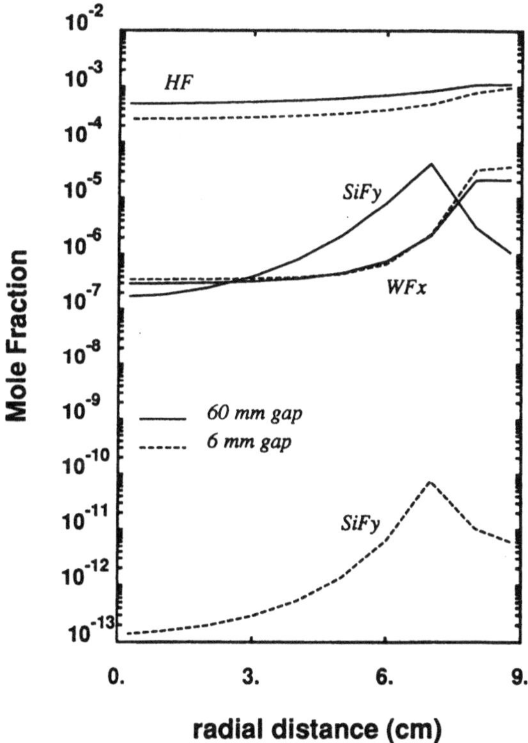

Figure 5.21: Concentration profiles across a wafer radius for two different reactor designs. For the smaller gap the HF and SiF_y concentrations are efficiently reduced.

completely different consequences of the design modification on selectivity loss. If we favor the SiF_y model, we would expect an extreme improvement in selectivity, whereas with the WF_x model we expect no influence.

In this context, we feel that the problems with selectivity loss, which still prevail in the semiconductor industry are a very strong argument for the WF_x model. At least, there must be a mechanism - maybe parallel to other possibilities - which cannot be removed just by blowing the HF away, otherwise reliable selective tungsten processes would have appeared already years ago.

5.4　Discussion

In our 2-D numerical simulation of selective tungsten deposition included were not only the concentration distributions of the main reaction species H_2 ,WF_6, and HF, but also reactive intermediates WF_x, and SiF_y, which are considered responsible for selectivity loss.

Basic experimental features like the time dependence of the nucleation density and the radial distribution of nuclei across the wafer are well reproduced by the model. However, the qualitative differences between the two chemical selectivity models studied are not significant enough to allow a decision between the two models merely from the comparison of experimental data with simulation results.

With both of the investigated selectivity models the simulation can be used for an evaluation of reactor design variation with respect to improved selectivity. Our calculations indicate that it is possible to reduce the HF concentration by changes in the reactor design, but whether this helps to improve the selectivity or not depends on the chemical model assumed for selectivity loss. Thus we conclude that the simulation model presented in this paper seems to be well suited to support the design and optimization of selective tungsten CVD reactors and processes, but it is important to have more experimental material about the chemical species responsible for selectivity loss.

References Chapter 5

[5.1] C.M. McConica, (1986), "A Model for Tungsten Nucleation on Oxide", in: R.S. Blewer (ed.), "Tungsten and other refractory metals for VLSI Application", Materials Research Society, Pittsburgh, PA p. 51.

[5.2] J.R. Creighton (1989), "A Mechanism of Selectivity Loss during Tungsten CVD", *J. Electrochem Soc.* **136** p.271

[5.3   C.M. McConica and K. Krishnamani (1986), "The Kinetics of LPCVD Tungsten Deposition in a Single Wafer Reactor", *J. Electrochem. Soc.* **133** p.2542.

[5.4] R.W.Pattee, C.M. McConica and K. Baughman (1988), "Polyimide Film Properties and Selective LPCVD of Tungsten on Polyimide", *J. Electrochem. Soc.* **135** p.1477.

[5.5] L.F.T. Kwakman, W.J.C. Vermeulen, E.H.A. Granneman, and M.L.Hitchman (1987), "A Quantitative Analysis of Reaction Byproducts on Selectivity during selective deposition of Tungsten", in: E.K. Broadbent (ed.), "Tungsten and other refractory metals for VLSI Applications II", Materials Research Society, Pittsburgh, PA p. 141

[5.6] H.Körner (1989), "Selective low pressure chemical vapor deposition of Tungsten", *Thin Solid Films*, **175** pp.55-60

[5.7] D.J. Bradbury, J.E.Turner, K. Nauka, and K.Y. Chiu (1991), "Selective CVD Tungsten as an alternative to blanket Tungsten for Submicron Plug Applications on VLSI Circuits", IEDM 91, Digest of Technical Papers, p.273, Washington 1991

[5.8] I Hirase, T. Sumiya, M.Schack, S. Ukishima, D. Rufin, M. Shishikura, M. Matsuura, and A. Ito (1987), "The effects of impurities and byproducts on selective deposition of Tungsten", in: E.K. Broadbent (ed.), "Tungsten and other refractory metals for VLSI Applications II",Materials Research Society, Pittsburgh, PA p.133

[5.9] Rikhit Arora and Richard Pollard (1991), "A Mathematical Model for Chemical Vapor Deposition Processes Influenced

by Surface Reaction Kinetics: Application to Low-Pressure
Deposition of Tungsten", *J. Electrochem. Soc.* **138** p.1523.

[5.10] H.I. Rosten and D.B. Spalding (1987), The PHOENICS
Beginner's Guide, CHAM TR/100, Wimbledon.

[5.11] G. Wahl (1985), "The Role of Thermal diffusion in CVD
Processes", in: J.O. Carlsson (ed), Proc. of the fifth
European Conf. on CVD, les editions de physique, Les Ulis
Cedex, France p.88

[5.12] Kenneth K. Kuo (1986), Principles of Combustion, p.669,
John Wiley & sons, New York

[5.13] R.B. Bird, W.E. Stewart, E.N. Lightfoot (1960), Transport
Phenomena, John Wiley & sons New York.

[5.14] R.C. Reid, J.M. Prausnitz, B.R. Poling (1987), The
Properties of Gases and Liquids, fourth edition,
Mc Graw-Hill, New York.

[5.15] E.K. Broadbent and C.L. Ramiller (1984), "Selective low
pressure chemical vapor deposition of tungsten", *Journ.
Electrochem. Soc.* **131** p.1427.

Chapter 6

Conclusions

Due to the worldwide efforts by many researchers, there has been an impressive evolution in the development of numerical simulation models for CVD chemistry and hydrodynamics in the last ten years. At present, major progress is being made in the integration of detailed chemistry models and multidimensional hydrodynamical models and in the application of these models to the study and optimization of realistic reactors and processes. In this book the state of the art has been described for the particular case of tungsten deposition. The model that has been developed and described in this study combines a full hydrodynamics model with detailed models for multicomponent transport phenomena and multiple chemical reactions. Such models have now reached a point where they can be used successfully in reactor design and process optimization, as has been illustrated here.

The rigorous approach described in this work makes sure that the same methods can be applied to different processes and reactor geometries. However, we must always be aware of the fact that all chemical and physical models are only approximative descriptions of a much more complicated reality, and hence some precautiosn concerning the limits of their validity are necessary.

The transport models,which were described in chapter 2, model the reality in a CVD reactor in a very good approximation as long as all distances of importance are a factor 100 or more larger than the molecular mean free path. In cases where this assumption is not fulfilled one can still produce reasonable results as was shown for the temperature drop across the submillimeter gap between wafer and susceptor in chapter 3, or for the step coverage in micron scale features of chapter 4. Nevertheless these first order corrections are of far less general validity and will require new considerations in every application.

The temperature modeling described in chapter 3, although very rigorous and generally applicable, requires experimental values for the emissivities of all the internal reactor surfaces. Unfortunately, these values are not material constants but depend on the mechanical treatment of the surface, so that no rigorous *a priori* calculations are possible without an experimental surface characterization.

Chemical modeling is based on pure empirics in our approach. The reaction pathways, reactive intermediates and rate constants cannot be obtained self-consistently from the present model and must be supplied seperately either from models with higher degree of sophistication (e.g. RRKM) or from experimental investigations.Although insufficient, part of these data is available for the tungsten deposition process studied in this book. For other CVD chemistries, except for the well studied silicon deposition process, the chemical knowledge is still at a much more rudimentary level.

From the limitations found so far, the important challenges for further developments in the near future can be deduced. In particular, the insufficient knowledge of the reaction pathways and kinetics is an important bottleneck for the future evolution and application of CVD simulation models. As indicated in this book, diagnostic kinetic experiments should be performed together with numerical simulations of the transport phenomena in order to accurately unravel the chemistry of the relevant CVD processes. Moreover, the tendency to operate CVD processes at very low pressures necessitates the development of non-continuum models for the transport phenomena in CVD reactors. Finally, in order to actually contribute to the development of better processes and reactors, and to reduced costs for research and development, numerical CVD simulation models should become more easily accessible for non-expert engineers working on the design and development of new reactors and processes. For this purpose, flexible and user-friendly computer codes have to be further developed. This study has explored many of the aspects that should be included in such codes.